THE LITTLE BOOK OF STRING THEORY

Books in the *SCIENCE ESSENTIALS* series bring cutting-edge science to a general audience. The series provides the foundation for a better understanding of the scientific and technical advances changing our world. In each volume, a prominent scientist—chosen by an advisory board of National Academy of Science members—conveys in clear prose the fundamental knowledge underlying a rapidly evolving field of scientific endeavor.

The Great Brain Debate: Nature or Nurture,
by John Dowling

Memory: The Key to Consciousness,
by Richard F. Thompson and Stephen Madigan

The Faces of Terrorism: Social and Psychological Dimensions,
by Neil J. Smelser

The Mystery of the Missing Antimatter,
by Helen R. Quinn and Yossi Nir

The Long Thaw: How Humans Are Changing the
Next 100,000 Years of Earth's Climate,
by David Archer

The Medea Hypothesis: Is Life on Earth Ultimately Self-Destructive?
by Peter Ward

How to Find a Habitable Planet,
by James Kasting

The Little Book of String Theory,
by Steven S. Gubser

the

LITTLE BOOK

of

STRING THEORY

STEVEN S. GUBSER

PRINCETON UNIVERSITY PRESS   PRINCETON AND OXFORD

Copyright 2010 © by Steven S. Gubser

Requests for permission to reproduce material from this work
should be sent to Permissions, Princeton University Press

Published by Princeton University Press,
41 William Street, Princeton, New Jersey 08540

In the United Kingdom: Princeton University Press,
6 Oxford Street, Woodstock, Oxfordshire OX20 1TW

All Rights Reserved

Library of Congress Cataloging-in-Publication Data

Gubser, Steven Scott, 1972-
The little book of string theory / Steven S. Gubser.
p. cm. — (Science essentials)
Includes index.
ISBN 978-0-691-14289-0 (cloth : alk. paper)
1. String models—Popular works. I. Title.
QC794.6.S85G83 2010
539.7'258—dc22
2009022871

British Library Cataloging-in-Publication Data is available

This book has been composed in Bembo

Printed on acid-free paper. ∞

press.princeton.edu

Printed in the United States of America

10 9 8 7 6 5 4 3 2 1

CONTENTS

To my father

INTRODUCTION

STRING THEORY IS A MYSTERY. IT'S SUPPOSED TO BE THE THE-
ory of everything. But it hasn't been verified experimentally. And it's so esoteric. It's all about extra dimensions,
quantum fluctuations, and black holes. How can that be the
world? Why can't everything be simpler?

String theory is a mystery. Its practitioners (of which I am
one) admit they don't understand the theory. But calculation
after calculation yields unexpectedly beautiful, connected
results. One gets a sense of inevitability from studying string
theory. How can this not be the world? How can such deep
truths fail to connect to reality?

String theory is a mystery. It draws many talented graduate students away from other fascinating topics, like superconductivity, that already have industrial applications. It
attracts media attention like few other fields in science. And
it has vociferous detractors who deplore the spread of its
influence and dismiss its achievements as unrelated to empirical science.

Briefly, the claim of string theory is that the fundamental
objects that make up all matter are not particles, but strings.
Strings are like little rubber bands, but very thin and very
strong. An electron is supposed to be actually a string, vibrating and rotating on a length scale too small for us to probe
even with the most advanced particle accelerators to date. In

2

some versions of string theory, an electron is a closed loop of string. In others, it is a segment of string, with two endpoints.

Let's take a brief tour of the historical development of string theory.

String theory is sometimes described as a theory that was invented backwards. Backwards means that people had pieces of it quite well worked out without understanding the deep meaning of their results. First, in 1968, came a beautiful formula describing how strings bounce off one another. The formula was proposed before anyone realized that strings had anything to do with it. Math is funny that way. Formulas can sometimes be manipulated, checked, and extended without being deeply understood. Deep understanding did follow in this case, though, including the insight that string theory included gravity as described by the theory of general relativity.

In the 1970s and early '80s, string theory teetered on the brink of oblivion. It didn't seem to work for its original purpose, which was the description of nuclear forces. While it incorporated quantum mechanics, it seemed likely to have a subtle inconsistency called an anomaly. An example of an anomaly is that if there were particles similar to neutrinos, but electrically charged, then certain types of gravitational fields could spontaneously create electric charge. That's bad because quantum mechanics needs the universe to maintain a strict balance between negative charges, like electrons, and positive charges, like protons. So it was a big relief when, in 1984, it was shown that string theory was free of anomalies. It was then perceived as a viable candidate to describe the universe.

This apparently technical result started the "first superstring revolution": a period of frantic activity and dramatic advances, which nevertheless fell short of its stated goal, to produce a theory of everything. I was a kid when it got going, and I lived close to the Aspen Center for Physics, a hotbed of activity. I remember people muttering about whether superstring theory might be tested at the Superconducting Super Collider, and I wondered what was so super about it all. Well, superstrings are strings with the special property of supersymmetry. And what might supersymmetry be? I'll try to tell you more clearly later in this book, but for now, let's settle for two very partial statements. First: Supersymmetry relates particles with different spins. The spin of a particle is like the spin of a top, but unlike a top, a particle can never stop spinning. Second: Supersymmetric string theories are the string theories that we understand the best. Whereas non-supersymmetric string theories require 26 dimensions, supersymmetric ones only require ten. Naturally, one has to admit that even ten dimensions is six too many, because we perceive only three of space and one of time. Part of making string theory into a theory of the real world is somehow getting rid of those extra dimensions, or finding some useful role for them.

For the rest of the 1980s, string theorists raced furiously to uncover the theory of everything. But they didn't understand enough about string theory. It turns out that strings are not the whole story. The theory also requires the existence of branes: objects that extend in several dimensions. The simplest brane is a membrane. Like the surface of a drum, a membrane extends in two spatial dimensions. It is a surface that can vibrate. There are also 3-branes, which can fill the three dimensions of space that we experience and vibrate in the additional dimensions that string theory requires. There can also be 4-branes, 5-branes, and so on up to 9-branes. All of this starts to sound like a lot to swallow, but there are solid reasons to believe that you can't make sense of string theory without all these branes included. Some of these reasons have

4

to do with "string dualities." A duality is a relation between two apparently different objects, or two apparently different viewpoints. A simplistic example is a checkerboard. One view is that it's a red board with black squares. Another view is that it's a black board with red squares. Both viewpoints (made suitably precise) provide an adequate description of what a checkerboard looks like. They're different, but related under the interchange of red and black.

The middle 1990s saw a second superstring revolution, based on the emerging understanding of string dualities and the role of branes. Again, efforts were made to parlay this new understanding into a theoretical framework that would qualify as a theory of everything. "Everything" here means all the aspects of fundamental physics we understand and have tested. Gravity is part of fundamental physics. So are electromagnetism and nuclear forces. So are the particles, like electrons, protons, and neutrons, from which all atoms are made. While string theory constructions are known that reproduce the broad outlines of what we know, there are some persistent difficulties in arriving at a fully viable theory. At the same time, the more we learn about string theory, the more we realize we don't know. So it seems like a third superstring revolution is needed. But there hasn't been one yet. Instead, what is happening is that string theorists are trying to make do with their existing level of understanding to make partial statements about what string theory might say about experiments both current and imminent. The most vigorous efforts along these lines aim to connect string theory with high-energy collisions of protons or heavy ions. The connections we hope for will probably hinge on the ideas of supersymmetry, or extra dimensions, or black hole horizons, or maybe all three at once.

INTRODUCTION

Now that we're up to the modern day, let's detour to consider the two types of collisions I just mentioned.

Proton collisions will soon be the main focus of experimental high-energy physics, thanks to a big experimental facility near Geneva called the Large Hadron Collider (LHC). The LHC will accelerate protons in counter-rotating beams and slam them together in head-on collisions near the speed of light. This type of collision is chaotic and uncontrolled. What experimentalists will look for is the rare event where a collision produces an extremely massive, unstable particle. One such particle—still hypothetical—is called the Higgs boson, and it is believed to be responsible for the mass of the electron. Supersymmetry predicts many other particles, and if they are discovered, it would be clear evidence that string theory is on the right track. There is also a remote possibility that proton-proton collisions will produce tiny black holes whose subsequent decay could be observed.

In heavy ion collisions, a gold or lead atom is stripped of all its electrons and whirled around the same machine that carries out proton-proton collisions. When heavy ions collide head-on, it is even more chaotic than a proton-proton collision. It's believed that protons and neutrons melt into their constituent quarks and gluons. The quarks and gluons then form a fluid, which expands, cools, and eventually freezes back into the particles that are then observed by the detectors. This fluid is called the quark-gluon plasma. The connection with string theory hinges on comparing the quark-gluon plasma to a black hole. Strangely, the kind of black hole that could be dual to the quark-gluon plasma is not in the four dimensions of our everyday experience, but in a five-dimensional curved spacetime.

It should be emphasized that string theory's connections to the real world are speculative. Supersymmetry might simply

6

not be there. The quark-gluon plasma produced at the LHC may really not behave much like a five-dimensional black hole. What is exciting is that string theorists are placing their bets, along with theorists of other stripes, and holding their breaths for experimental discoveries that may vindicate or shatter their hopes.

This book builds up to some of the core ideas of modern string theory, including further discussion of its potential applications to collider physics. String theory rests on two foundations: quantum mechanics and the theory of relativity. From those foundations it reaches out in a multitude of directions, and it's hard to do justice to even a small fraction of them. The topics discussed in this book represent a slice across string theory that largely avoids its more mathematical side. The choice of topics also reflects my preferences and prejudices, and probably even the limits of my understanding of the subject.

Another choice I've made in writing this book is to discuss physics but not physicists. That is, I'm going to do my best to tell you what string theory is about, but I'm not going to tell you about the people who figured it all out (although I will say up front that mostly it wasn't me). To illustrate the difficulties of doing a proper job of attributing ideas to people, let's start by asking who figured out relativity. It was Albert Einstein, right? Yes—but if we just stop with that one name, we're missing a lot. Hendrik Lorentz and Henri Poincaré did important work that predated Einstein; Hermann Minkowski introduced a crucially important mathematical framework; David Hilbert independently figured out a key building block of general relativity; and there are several more important early figures like James Clerk Maxwell, George FitzGerald, and Joseph Larmor who deserve mention, as well as later pioneers like John Wheeler and Subrahmanyan Chandrasekhar. The development of quantum mechanics is considerably more intricate, as there is no single figure like Einstein whose contributions tower above all others. Rather, there is a fascinating and heterogeneous group, including Max Planck, Einstein, Ernest Rutherford, Niels Bohr, Louis de Broglie, Werner Heisenberg, Erwin Schrödinger, Paul Dirac, Wolfgang Pauli, Pascual Jordan, and John von Neumann, who contributed in essential ways—and sometimes famously disagreed with one another. It would be an even more ambitious project to properly assign credit for the vast swath of ideas that is string theory. My feeling is that an attempt to do so would actually detract from my primary aim, which is to convey the ideas themselves.

The aim of the first three chapters of this book is to introduce ideas that are crucial to the understanding of string theory, but that are not properly part of it. These ideas— energy, quantum mechanics, and general relativity—are more important (so far) than string theory itself, because we know that they describe the real world. Chapter 4, where I introduce string theory, is thus a step into the unknown. While I attempt in chapters 4, 5, and 6 to make string theory, D-branes, and string dualities seem as reasonable and well motivated as I can, the fact remains that they are unverified as descriptions of the real world. Chapters 7 and 8 are devoted to modern attempts to relate string theory to experiments involving high-energy particle collisions. Supersymmetry, string dualities, and black holes in a fifth dimension all figure in string theorists' attempts to understand what is happening, and what will happen, in particle accelerators.

INTRODUCTION

In various places in this book, I quote numerical values for physical quantities: things like the energy released in nuclear fission or the amount of time dilation experienced by an Olympic sprinter. Part of why I do this is that physics is a quantitative science, where the numerical sizes of things matter. However, to a physicist, what's usually most interesting is the approximate size, or order of magnitude, of a physical quantity. So, for example, I remark that the time dilation experienced by an Olympic sprinter is about a part in 10^{15} even though a more precise estimate, based on a speed of 10 m/s, is a part in 1.8×10^{15}. Readers wishing to see more precise, explicit, and/or extended versions of the calculations I describe in the book can visit this website: http://press.princeton.edu/titles/9133.html.

Where is string theory going? String theory promises to unify gravity and quantum mechanics. It promises to provide a single theory encompassing all the forces of nature. It promises a new understanding of time, space, and additional dimensions as yet undiscovered. It promises to relate ideas as seemingly distant as black holes and the quark–gluon plasma. Truly it is a "promising" theory!

How can string theorists ever deliver on the promise of their field? The fact is, much has been delivered. String theory does provide an elegant chain of reasoning starting with quantum mechanics and ending with general relativity. I'll describe the framework of this reasoning in chapter 4. String theory does provide a provisional picture of how to describe all the forces of nature. I'll outline this picture in chapter 7 and tell you some of the difficulties with making it more precise. And as I'll explain in chapter 8, string theory calculations are already being compared to data from heavy ion collisions.

I don't aim to settle any debates about string theory in this book, but I'll go so far as to say that I think a lot of the disagreement is about points of view. When a noteworthy result comes out of string theory, a proponent of the theory might say, "That was fantastic! But it would be so much better if only we could do thus-and-such." At the same time, a critic might say, "That was pathetic! If only they had done thus-and-such, I might be impressed." In the end, the proponents and the critics (at least, the more serious and informed members of each camp) are not that far apart on matters of substance. Everyone agrees that there are some deep mysteries in fundamental physics. Nearly everyone agrees that string theorists have mounted serious attempts to solve them. And surely it can be agreed that much of string theory's promise has yet to be delivered upon.

INTRODUCTION

Chapter ONE

ENERGY

THE AIM OF THIS CHAPTER IS TO PRESENT THE MOST FAMOUS equation of physics: $E = mc^2$. This equation underlies nuclear power and the atom bomb. It says that if you convert one pound of matter entirely into energy, you could keep the lights on in a million American households for a year. $E = mc^2$ also underlies much of string theory. In particular, as we'll discuss in chapter 4, the mass of a vibrating string receives contributions from its vibrational energy.

What's strange about the equation $E = mc^2$ is that it relates things you usually don't think of as related. E is for energy, like the kilowatt-hours you pay your electric company for each month; m is for mass, like a pound of flour; c is for the speed of light, which is 299,792,458 meters per second, or (approximately) 186,282 miles per second. So the first task is to understand what physicists call "dimensionful quantities," like length, mass, time, and speed. Then we'll get back to $E = mc^2$ itself. Along the way, I'll introduce metric units, like meters and kilograms; scientific notation for big numbers; and a bit of nuclear physics. Although it's not necessary to understand nuclear physics in order to grasp string theory, it provides a good context for discussing $E = mc^2$. And in chapter 8, I will come back and explain efforts to use string theory to better understand aspects of modern nuclear physics.

Length, mass, time, and speed

Length is the easiest of all dimensionful quantities. It's what you measure with a ruler. Physicists generally insist on using the metric system, so I'll start doing that now. A meter is about 39.37 inches. A kilometer is 1000 meters, which is about 0.6214 miles.

Time is regarded as an additional dimension by physicists. We perceive four dimensions total: three of space and one of time. Time is different from space. You can move any direction you want in space, but you can't move backward in time. In fact, you can't really "move" in time at all. Seconds tick by no matter what you do. At least, that's our everyday experience. But it's actually not that simple. If you run in a circle really fast while a friend stands still, time as you experience it will go by less quickly. If you and your friend both wear stopwatches, yours will show less time elapsed than your friend's. This effect, called time dilation, is imperceptibly small unless the speed with which you run is comparable to the speed of light.

Mass measures an amount of matter. We're used to thinking of mass as the same as weight, but it's not. Weight has to do with gravitational pull. If you're in outer space, you're weightless, but your mass hasn't changed. Most of the mass in everyday objects is in protons and neutrons, and a little bit more is in electrons. Quoting the mass of an everyday object basically comes down to saying how many nucleons are in it. A nucleon is either a proton or a neutron. My mass is about 75 kilograms. Rounding up a bit, that's about 50,000,000,000, 000,000,000,000,000,000 nucleons. It's hard to keep track of such big numbers. There are so many digits that you can't easily count them up. So people resort to what's called scientific notation: instead of writing out all the digits like I did before, you would say that I have about 5×10^{28} nucleons in me. The 28 means that there are 28 zeroes after the 5. Let's practice a bit more. A million could be written as 1×10^6, or, more simply, as 10^6. The U.S. national debt, currently about $10,000,000,000,000, can be conveniently expressed as 10^{13} dollars. Now, if only I had a dime for every nucleon in me . . .

Let's get back to dimensionful quantities in physics. Speed is a conversion factor between length and time. Suppose you can run 10 meters per second. That's fast for a person—really fast. In 10 seconds you can go 100 meters. You wouldn't win an Olympic gold with that time, but you'd be close. Suppose you could keep up your speed of 10 meters per second over any distance. How long would it take to go one kilometer? Let's work it out. One kilometer is ten times 100 meters. You can do the 100-meter dash in 10 seconds flat. So you can run a kilometer in 100 seconds. You could run a mile in 161 seconds, which is 2 minutes and 41 seconds. No one can do that, because no one can keep up a 10 m/s pace for that long.

Suppose you could, though. Would you be able to notice the time dilation effect I described earlier? Not even close. Time would run a little slower for you while you were pounding out your 2:41 mile, but slower only by one part in about 10^{15} (that's a part in 1,000,000,000,000,000, or a thousand million million). In order to get a big effect, you would have to be moving much, much faster. Particles whirling around modern accelerators experience tremendous time dilation. Time for them runs about 1000 times slower than for a proton at rest. The exact figure depends on the particle accelerator in question.

The speed of light is an awkward conversion factor for everyday use because it's so big. Light can go all the way around the equator of the Earth in about 0.1 seconds. That's part of why an American can hold a conversation by telephone with someone in India and not notice much time lag. Light is more useful when you're thinking of really big distances. The distance to the moon is equivalent to about 1.3 seconds. You could say that the moon is 1.3 light-seconds away from us. The distance to the sun is about 500 light-seconds.

A light-year is an even bigger distance: it's the distance that light travels in a year. The Milky Way is about 100,000 light-years across. The known universe is about 14 billion light-years across. That's about 1.3×10^{26} meters.

$$E = mc^2$$

The formula $E = mc^2$ is a conversion between mass and energy. It works a lot like the conversion between time and distance that we just discussed. But just what is energy? The question is hard to answer because there are so many forms of energy. Motion is energy. Electricity is energy. Heat is energy. Light is energy. Any of these things can be converted into any other. For example, a lightbulb converts electricity into heat and light, and an electric generator converts motion into electricity. A fundamental principle of physics is that total energy is conserved, even as its form may change. In order to make this principle meaningful, one has to have

CHAPTER ONE

ways of quantifying different forms of energy that can be converted into one another.

A good place to start is the energy of motion, also called kinetic energy. The conversion formula is $K = \frac{1}{2}mv^2$, where K is the kinetic energy, m is the mass, and v is the speed. Imagine yourself again as an Olympic sprinter. Through a tremendous physical effort, you can get yourself going at $v = 10$ meters per second. But this is much slower than the speed of light. Consequently, your kinetic energy is much less than the energy E in $E = mc^2$. What does this mean?

It helps to know that $E = mc^2$ describes "rest energy." Rest energy is the energy in matter when it is not moving. When you run, you're converting a little bit of your rest energy into kinetic energy. A very little bit, actually: roughly one part in 10^{15}. It's no accident that this same number, one part in 10^{15}, characterizes the amount of time dilation you experience when you run. Special relativity includes a precise relation between time dilation and kinetic energy. It says, for example, that if something is moving fast enough to double its energy, then its time runs half as fast as if it weren't moving.

It's frustrating to think that you have all this rest energy in you, and all you can call up with your best efforts is a tiny fraction, one part in 10^{15}. How might we call up a greater fraction of the rest energy in matter? The best answer we know of is nuclear energy.

Our understanding of nuclear energy rests squarely on $E = mc^2$. Here is a brief synopsis. Atomic nuclei are made up of protons and neutrons. A hydrogen nucleus is just a proton. A helium nucleus comprises two protons and two neutrons, bound tightly together. What I mean by tightly bound is that it takes a lot of energy to split a helium nucleus. Some nuclei are much easier to split. An example is uranium-235, which is made of 92 protons and 143 neutrons. It is quite easy to break a uranium-235 nucleus into several pieces. For instance, if you hit a uranium-235 nucleus with a neutron, it can split into a krypton nucleus, a barium nucleus, three neutrons, and energy. This is an example of fission. We could write the reaction briefly as

$$U + n \rightarrow Kr + Ba + 3n + \text{Energy},$$

where we understand that U stands for uranium-235, Kr stands for krypton, Ba stands for barium, and n stands for neutron. (By the way, I'm careful always to say uranium-235 because there's another type of uranium, made of 238 nucleons, that is far more common, and also harder to split.)

$E = mc^2$ allows you to calculate the amount of energy that is released in terms of the masses of all the participants in the fission reaction. It turns out that the ingredients (one uranium-235 nucleus plus one neutron) outweigh the products (a krypton atom, a barium atom, and three neutrons) by about a fifth of the mass of a proton. It is this tiny increment of mass that we feed into $E = mc^2$ to determine the amount of energy released. Tiny as it seems, a fifth of the mass of a proton is almost a tenth of a percent of the mass of a uranium-235 atom: one part in a thousand. So the energy released is about a thousandth of the rest energy in a uranium-235 nucleus. This still may not seem like much, but it's roughly a trillion times bigger as a fraction of rest energy than the fraction that an Olympic sprinter can call up in the form of kinetic energy.

I still haven't explained where the energy released in nuclear fission comes from. The number of nucleons doesn't change: there are 236 of them before and after fission. And yet the ingredients have more mass than the products. So this is an important exception to the rule that mass is essentially a count of nucleons. The point is that the nucleons in the krypton and barium nuclei are bound more tightly than they were in the uranium-235 nucleus. Tighter binding means less mass. The loosely bound uranium-235 nucleus has a little extra mass, just waiting to be released as energy. To put it in a nutshell: Nuclear fission releases energy as protons and neutrons settle into slightly more compact arrangements.

One of the projects of modern nuclear physics is to figure out what happens when heavy nuclei like uranium-235 undergo far more violent reactions than the fission reaction I described. For reasons I won't go into, experimentalists prefer to work with gold instead of uranium. When two gold nuclei are slammed into one another at nearly the speed of light, they are utterly destroyed. Almost all the nucleons break up. In chapter 8, I will tell you more about the dense, hot state of matter that forms in such a reaction.

In summary, $E = mc^2$ says that the amount of rest energy in something depends only on its mass, because the speed of light is a known constant. It's easier to get some of that energy out of uranium-235 than most other forms of matter. But fundamentally, rest energy is in all forms of matter equally: rocks, air, water, trees, and people.

Before going on to quantum mechanics, let's pause to put $E = mc^2$ in a broader intellectual context. It is part of special relativity, which is the study of how motion affects measurements of time and space. Special relativity is subsumed in general relativity, which also encompasses gravity and curved spacetime. String theory subsumes both general relativity and quantum mechanics. In particular, string theory includes the relation $E = mc^2$. Strings, branes, and black

18

holes all obey this relation. For example, in chapter 5 I'll discuss how the mass of a brane can receive contributions from thermal energy on the brane. It wouldn't be right to say that $E = mc^2$ follows from string theory. But it fits, seemingly inextricably, with other aspects of string theory's mathematical framework.

CHAPTER ONE

Chapter T W O

QUANTUM MECHANICS

AFTER I GOT MY BACHELOR'S DEGREE IN PHYSICS, I SPENT A year at Cambridge University studying math and physics. Cambridge is a place of green lawns and grey skies, with an immense, historical weight of genteel scholarship. I was a member of St. John's College, which is about five hundred years old. I particularly remember playing a fine piano located in one of the upper floors of the first court—one of the oldest bits of the college. Among the pieces I played was Chopin's Fantasie-Impromptu. The main section has a persistent four-against-three cross rhythm. Both hands play in even tempo, but you play four notes with the right hand for every three notes in the left hand. The combination gives the composition an ethereal, liquid sound.

It's a beautiful piece of music. And it makes me think about quantum mechanics. To explain why, I will introduce some concepts of quantum mechanics, but I won't try to explain them completely. Instead, I will try to explain how they combine into a structure that is, to me, reminiscent of music like the Fantasie-Impromptu. In quantum mechanics, every motion is possible, but there are some that are preferred. These preferred motions are called quantum states. They have definite frequencies. A frequency is the number of times per second that something cycles or repeats. In the Fantasie-Impromptu, the patterns of the right hand have a faster frequency, and the patterns of the left hand have a slower frequency, in the ratio four to three. In quantum systems, the thing that is cycling is more abstract: technically, it's the phase of the wave function. You can think of the phase of the wave function as similar to the second hand of a clock. The second hand goes around and around, once per minute. The phase is doing the same thing, cycling around at some much faster frequency. This rapid cycling characterizes the energy of the system in a way that I'll discuss in more detail later.

Simple quantum systems, like the hydrogen atom, have frequencies that stand in simple ratios with one another. For example, the phase of one quantum state might cycle nine times while another cycles four times. That's a lot like the four-against-three cross rhythm of the Fantasie-Impromptu. But the frequencies in quantum mechanics are usually a lot faster. For example, in a hydrogen atom, characteristic frequencies are on the scale of 10^{15} oscillations or cycles per second. That's indeed a lot faster than the Fantasie-Impromptu, in which the right hand plays about 12 notes per second.

The rhythmic fascination of the Fantasie-Impromptu is hardly its greatest charm—at least, not when it's played rather better than I ever could. Its melody floats above a melancholy bass. The notes run together in a chromatic blur. The harmonies shift slowly, contrasting with the almost desultory flitting of the main theme. The subtle four-against-three rhythm provides just the backdrop for one of Chopin's more memorable compositions. Quantum mechanics is like this. Its underlying graininess, with quantum states at definite frequencies, blurs at larger scales into the colorful, intricate world of our experience. Those quantum frequencies leave an indelible mark on that world: for example, the orange light from a street lamp has a definite frequency, associated with a particular cross rhythm in sodium atoms. The frequency of the light is what makes it orange.

In the rest of this chapter, I'm going to focus on three aspects of quantum mechanics: the uncertainty principle, the hydrogen atom, and the photon. Along the way, we'll encounter energy in its new quantum mechanical guise, closely related to frequency. Analogies with music are apt for those aspects of quantum mechanics having to do with frequency. But as we'll see in the next section, quantum physics incorporates some other key ideas that are less readily compared with everyday experience.

Uncertainty

One of the cornerstones of quantum mechanics is the uncertainty principle. It says that a particle's position and momentum can never be simultaneously measured. That's an oversimplification, so let me try to do better. Any measurement of position will have some uncertainty, call it Δx (pronounced "Delta x"). For instance, if you measure the length of a piece of wood with a tape measure, you can usually get it right to within 1/32 of an inch if you're careful. That's a little less than a millimeter. So for such a measurement, one would say $\Delta x \approx 1\text{mm}$: that is, "Delta x (the uncertainty) is approximately one millimeter." Despite the Greek letter Δ, the concept here is simple: A carpenter might call out to his

22

buddy, "Jim, this board is within a millimeter of two meters long." (Of course, I'm referring to a European carpenter, since the guys I've seen in the United States prefer feet and inches.) What the carpenter means is that the length of the board is $x = 2$ meters, with an uncertainty $\Delta x \approx 1\,\text{mm}$.

Momentum is familiar from everyday experience, but to be precise about it, it helps to consider collisions. If two things collide head-on and the impact stops them both completely, then they had equal momentum before the collision. If after the collision one is still moving in the direction it started, but slower, then it had larger momentum. There's a conversion formula from mass m to momentum p: $p = mv$. But let's not worry about the details just yet. The point is that momentum is something you can measure, and the measurement has some uncertainty, which we'll call Δp.

The uncertainty principle says $\Delta p \times \Delta x \geq h/4\pi$, where h is a quantity called Planck's constant and $\pi = 3.14159 \ldots$ is the familiar ratio of the circumference of a circle to its diameter. I would read this formula aloud as "Delta p times Delta x is no less than h over 4 Pi." Or, if you prefer, "The product of the uncertainties in a particle's momentum and position is never less than Planck's constant divided by 4 Pi." Now you can see why my original statement of the uncertainty principle was an oversimplification: You *can* measure position and momentum simultaneously, but the uncertainty in those two measurements can never be smaller than what the equation $\Delta p \times \Delta x \geq h/4\pi$ allows.

To understand an application of the uncertainty principle, think of capturing a particle in a trap whose size is Δx. The position of the particle is known within an uncertainty Δx if it is in the trap. The uncertainty principle then says is that it's impossible to know the momentum of the trapped particle

CHAPTER TWO

more precisely than a certain bound. Quantitatively, the uncertainty in the momentum, Δp, has to be large enough so that the inequality $\Delta p \times \Delta x \geq h/4\pi$ is satisfied. Atoms provide an example of all this, as we'll see in the next section. It's hard to give a more everyday example, because typical uncertainties Δx are much smaller than objects that you can hold in your hand. That's because Planck's constant is numerically very small. We'll encounter it again when we discuss photons, and I'll tell you then what its numerical value actually is.

The way you usually talk about the uncertainty principle is to discuss measurements of position and momentum. But it goes deeper than that. It is an intrinsic limitation on what position and momentum mean. Ultimately, position and momentum are not numbers. They are more complicated objects called operators, which I won't try to describe except to say that they are perfectly precise mathematical constructions—just more complicated than numbers. The uncertainty principle arises from the difference between numbers and operators. The quantity Δx is not just the uncertainty of a measurement; it is the irreducible uncertainty of the particle's position. What the uncertainty principle captures is not a lack of knowledge, but a fundamental fuzziness of the subatomic world.

The atom

The atom is made up of electrons moving around the atomic nucleus. As we've already discussed, the nucleus is made up of protons and neutrons. The simplest case to start with is hydrogen, where the nucleus is just a proton, and there's only one electron moving around it. The size of an atom is roughly 10^{-10} meters, also known as an angstrom. (Saying that an angstrom is 10^{-10} meters means that a meter is 10^{10}, or ten billion, angstroms.) The size of a nucleus is about a hundred thousand times smaller. When one says that an atom is about an angstrom across, it means the electron rarely goes further away from the nucleus than this. The uncertainty Δx in the position of the electron is about an angstrom, because from instant to instant it's impossible to say on which side of the nucleus the electron will find itself. The uncertainty principle then says that there's an uncertainty Δp in the momentum of the electron, satisfying $\Delta p \times \Delta x \geq h/4\pi$. The way this comes about is that the electron in the hydrogen atom has some average speed—about a hundredth of the speed of light—but which direction it is moving in changes from moment to moment and is fundamentally uncertain. The uncertainty in the momentum of the electron is essentially the momentum itself, because of this uncertainty in direction. The overall picture is that the electron is trapped by its attraction to the nucleus, but quantum mechanics does not permit the electron to rest in this trap. Instead, it wanders ceaselessly in a way that the mathematics of quantum mechanics describes. This insistent wandering is what gives the atom its size. If the electron were permitted to sit still, it would do so inside the nucleus, because it is attracted to the nucleus. Matter itself would then collapse to the density of the nucleus, which would be very uncomfortable! So the quantum wanderings of the electrons inside atoms are really a blessing.

Although the electron in a hydrogen atom has an uncertain position and an uncertain momentum, it has a definite energy. Actually, it has several possible energies. The way physicists describe the situation is to say that the electron's energy is "quantized." That means that it has to choose among a definite set of possibilities. To appreciate this strange state

CHAPTER TWO

of affairs, let's go back to kinetic energy in an everyday example. We learned about the conversion formula, $K = \frac{1}{2}mv^2$. Let's say we apply it to a car. By giving the car more and more gas, you can pick out whatever speed v that you want. However, if energy were quantized for a car, you wouldn't be able to do this. For example, you might be able to go 10 miles per hour, or 15, or 25, but not 11 or 12 or 12.5 miles per hour.

The quantized energy levels of the electron in hydrogen bring me back to analogies with music. I already introduced one such analogy: the cross-rhythms in the Fantasie-Impromptu. A steady rhythm is itself a frequency. Each quantized energy level in hydrogen corresponds to a different frequency. An electron can pick one of these levels. If it does, that's like having a single steady rhythm, like a metronome. But an electron can also choose to be partly in one energy level and partly in another. That's called a superposition. The Fantasie-Impromptu is a "superposition" of two different rhythms, one carried by the right hand and one by the left.

So far, I've told you that electrons in atoms have quantum mechanically uncertain position and momentum, but quantized energies. Isn't it strange that energies should be fixed to definite values when position and momentum cannot be fixed? To understand how this comes about, let's detour into another analogy with music. Think of a piano string. When struck, it vibrates with a definite frequency, or pitch. For example, A above middle C on a piano vibrates 440 times in a second. Often, physicists quote frequencies in terms of the hertz (abbreviated Hz), which is one cycle or oscillation per second. So A above middle C has frequency 440 Hz. That's much faster than the rhythms of the Fantasie-Impromptu, where, if you recall, the right hand plays about 12 notes in a second: a frequency of 12 Hz. But it's still much, much slower than a hydrogen atom's frequencies. Actually, the motion of the string is more complicated than a single vibration. There are overtones at higher frequencies. These overtones give a piano its characteristic sound.

This may seem distant from the quantum mechanical motion of an electron in a hydrogen atom. But in fact, it's closely related. The lowest energy of an electron in hydrogen is like the fundamental frequency of a piano string: 440 Hz for A above middle C. Oversimplifying a little, the frequency of an electron in its lowest energy state is about 3×10^{15} Hz. The other energies that are possible for an electron are like the overtones of a piano string.

The waves on a piano string and the quantum mechanical motions of an electron in a hydrogen atom are examples of standing waves. A standing wave is a vibration that doesn't go anywhere. The piano string is tied down at the ends, so its vibrations are confined to its length. The quantum mechanical motions of an electron in a hydrogen atom are confined to a much smaller space, little more than an angstrom across. The main idea behind the mathematics of quantum mechanics is to treat the motions of the electron as a wave. When the wave has a definite frequency, like the fundamental frequency of a piano string, the electron has a definite energy. But the position of the electron is never a definite number, because the wave describing it is everywhere in the atom at once, just as the vibration of a piano string is a vibration of the whole string at once. All you can say is that the electron is almost always within an angstrom of the nucleus.

Having learned that electrons are described by waves, you might ask: Waves on what? This is a hard question. One answer is that it doesn't seem to matter. Another is that there is an "electron field" permeating all of spacetime, and electrons

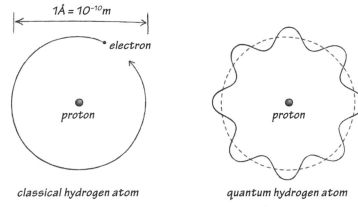

classical hydrogen atom   quantum hydrogen atom

Left: The classical picture of a hydrogen atom, where an electron orbits around a proton. Right: The quantum picture in terms of standing waves. Instead of following a definite orbit, the electron is represented as a standing wave. It is not at a definite position, but it does have a definite energy.

are excitations of it. The electron field is like the piano string, and the electron is like a vibration of the piano string.

Waves are not always confined to small spaces like the inside of an atom. For example, waves on the sea can travel many miles before breaking on the shore. There are examples of traveling waves in quantum mechanics: photons, for instance. But before we delve into photons, there's a technicality which I have to discuss because it relates to things that will come up again in later chapters. I quoted a frequency for an electron in hydrogen, and I remarked that I was oversimplifying. To explain how I was oversimplifying, I'll introduce one more formula: $E = h\nu$. Here E is energy, ν is frequency, and h the same Planck's constant that came up in the uncertainty principle. $E = h\nu$ is a wonderful formula, because it tells you what frequency really means: it's just energy in a new guise. But here's the trouble: There are different kinds

QUANTUM MECHANICS

of energy. The electron has rest energy. It has kinetic energy. And it has binding energy, which is the amount of energy it takes to knock the electron free of the proton. Which should you use in the formula $E = h\nu$? When I quoted the figure 3×10^{15} oscillations per second for hydrogen, I was using the kinetic energy plus the binding energy, with the rest energy excluded. But this was arbitrary. I could have included the rest energy if I felt like it. That means that frequency has some ambiguity in quantum mechanics, which seems awful.

Here's how the difficulties get resolved. You can ask what happens when an electron jumps from one energy level to another. If the electron jumps down in energy, then it sheds the excess energy by emitting a photon. The photon's energy is the difference between the electron's energy before it jumped and after. Now it doesn't matter whether you include the rest energy of the electron, because all you care about is the energy difference before and after the electron jumped levels. The proper use of the formula $E = h\nu$ is to set E equal to the energy of the photon. Then ν is the frequency of the photon, which is a definite number with no ambiguities. There's just one thing left to settle: Exactly what does the frequency of a photon mean? This is what I want to explain next.

The photon

For centuries, a debate raged in physics. Is light a particle, or is it a wave? Quantum mechanics resolved the debate in an unexpected way: it's both.

To appreciate the wavelike characteristics of light, imagine an electron who decides to go sunbathing in a laser beam. A laser beam is a steady, coherent, intense beam of

CHAPTER TWO

light. Here's the key point: When the electron steps into the laser beam, it pulls him first to one side, then the other, back and forth at some frequency. That frequency is the one that enters into the equation $E = h\nu$. Visible light has a frequency somewhat less than 10^{15} oscillations per second.

This analogy is fanciful, but it's easy to give a much more practical example. Radio waves are really the same thing as light, just with a much smaller frequency. FM radio waves have a frequency of about 10^8 oscillations per second, or 10^8 Hz. One of the most popular stations where I live is New Jersey 101.5, which broadcasts at 101.5 megahertz. One megahertz is a million hertz, or 10^6 Hz. So 100 megahertz is 10^8 Hz. Thus 101.5 megahertz is just a smidgen over 10^8 oscillations per second. An FM radio is constructed so that electrons inside it can oscillate at just about this frequency. When you tune a radio, what you are adjusting is the frequency at which the electrons inside its circuitry prefer to oscillate. Much like our sunbathing electron, the electrons inside the radio soak up the radio waves washing over the radio.

Another analogy that might help is a buoy in the ocean. Typically, a buoy is attached by a chain to an anchor at the bottom of the ocean so that it doesn't get washed away by the waves and currents. The way it responds to waves is to bob up and down, staying at the surface of the water. This is similar to the way the sunbathing electron responds to the laser beam. There's actually more to the story of the sunbathing electron: eventually it gets pushed in the direction of the laser beam, unless it is somehow tied down like the buoy.

So far, my explanation has focused on the wavelike properties of light. In what ways does it behave like a particle? There's a famous phenomenon called the photoelectric effect which provides evidence that light really is composed

QUANTUM MECHANICS

of photons, each with energy $E = h\nu$. Here's how it works. If you shine light on a metal, you knock electrons out of it. With a clever experimental apparatus, you can detect those electrons and even measure their energy. The results of such measurements are consistent with the following explanation. Light, composed of many photons, delivers a series of little kicks to the metal. Each kick arises when a photon hits one of the electrons in the metal. Sometimes, if the photon has enough energy, it can kick the electron it hits clean out of the metal. According to the equation $E = h\nu$, higher frequency means higher energy. It's known that blue light has a frequency approximately 35% higher than red light. That means a blue photon has 35% more energy than a red photon. Suppose you use sodium to study the photoelectric effect. It so happens that red photons aren't quite energetic enough to knock electrons out of sodium. You don't see any electrons getting knocked out even if you make the red light quite bright. But blue photons, with their extra increment of energy, do have enough energy to knock electrons out of sodium. Even a very faint blue light will do the trick. What matters is not the brightness—which is related to how many photons there are—but the color of the light, which determines the energy of each photon.

The minimum frequency of light it takes to kick electrons out of sodium is 5.5×10^{14} oscillations per second, which means green light. The corresponding energy, according to the equation $E = h\nu$, is 2.3 electron volts. An electron volt is the amount of energy a single electron acquires when it is pushed on by a one-volt power supply. So the numerical value of Planck's constant is 2.3 electron volts divided by 5.5×10^{14} oscillations per second. That's usually quoted briefly as 4.1×10^{-15} electron–volt–seconds.

CHAPTER TWO

In summary, light behaves like a wave in many circumstances, and like a particle in others. This is called wave-particle duality. According to quantum mechanics, it's not just light that exhibits wave-particle duality: everything does.

Let's return to the hydrogen atom for a moment. I tried to explain in the last section how its quantized energy levels can be thought of as standing waves with definite frequencies. That's an example of how electrons behave like waves. But if you remember, I got hung up on how to explain just what the frequency meant. I introduced the formula $E = h\nu$, but then I ran into trouble on the question of whether to include the rest energy of the electron in E. With photons, there's no such difficulty. The frequency of light really means something tangible. It's the frequency that you tune a radio to receive. So when an electron jumps from one energy level to another, emitting a single photon in the process, you can use the frequency of the photon to judge unambiguously what the energy difference is between the two levels.

I hope that the discussion so far has given you a fairly good feeling for what photons are. Understanding them completely is remarkably difficult. The difficulties hinge on a concept called gauge symmetry, which I'll discuss at some length in chapter 5. In the rest of this section, let's explore how photons weave together concepts from special relativity and quantum mechanics.

The theory of relativity is based on the assumption that light in a vacuum always goes at the same speed (299,792,458 meters per second), and that nothing can go faster. Everyone who contemplates these claims is eventually struck by the thought that if you accelerated yourself to the speed of light and then fired a pistol in the direction you were moving, the bullet would then be moving faster than the speed of light. Right? Not so fast. The trouble is connected with time dilation. Remember how I remarked that time ran 1000 times slower for particles in modern particle accelerators? It's because they're moving close to the speed of light. If, instead of moving close to the speed of light, you move *at* the speed of light, then time stops completely. You'd never fire that pistol, because you'd never have time to pull the trigger.

It might seem that this argument leaves a little wiggle room. Maybe you could get yourself to within 10 m/s of the speed of light. Time would be moving as slow as molasses for you, but eventually you could squeeze off a shot from your pistol. When you did, the bullet would be moving relative to you quite a bit faster than 10 m/s, so surely it would exceed the speed of light. Right? Well, it just never works out that way. The faster you're going, the harder it is to get anything moving faster than you are. It's not because there's some kind of headwind blowing in your face: this could all be happening in outer space. It's because of the way time, length, and speed get tangled up in special relativity. Everything in relativity is cooked up in just such a way as to frustrate any attempt to go faster than light. Given the many successes of relativity in describing the world, most physicists are inclined to accept its main claim at face value: You just can't go faster than light.

Now, what about the additional claim that light always goes at the same speed in a vacuum? This claim can be tested experimentally, and it seems to be true, no matter what frequency of light you use. This means that there is quite a stark contrast between photons and other particles, like electrons or protons. Electrons and protons can be fast or slow. If they're fast, they have lots of energy. If they're slow, they have less energy. But an electron by itself never has less energy than its rest energy, $E = mc^2$. Likewise the energy of a proton by itself is never less than *its* rest energy. A photon's energy, however, is $E = h\nu$, and the frequency ν can be as big or as small as we like without changing the speed of the photon. In particular, there's no lower limit on the energy of a photon. What that must mean is that the rest energy of a photon is zero. If we use $E = mc^2$, then we conclude that a photon's mass must be zero. That's the crucial difference between a photon and most other particles: a photon has no mass.

It won't matter to future discussions in this book, but it's nice to know that it's only in a vacuum that light has a fixed speed. Light in fact does slow down when it's passing through matter. What I have in mind now is a very different situation than visible light hitting sodium: instead, I'm thinking of light passing through a transparent substance like water or glass. When passing through water, light slows down by a factor of about 1.33. When it's passing through glass, it can slow down by more, but never as much as a factor of 2. Diamond slows light down by a factor of 2.4. This large factor, together with the clarity of diamond, gives it its unique, winking brilliance.

Chapter T H R E E

GRAVITY AND BLACK HOLES

ONE FINE SUMMER DAY SOME YEARS AGO, I DROVE WITH MY father up to Grotto Wall, a popular climbing crag near Aspen, Colorado. We aimed to climb a classic moderate route called Twin Cracks. After we finished it without incident, I trotted out another idea: aid-climbing a harder route called Cryogenics. Aid-climbing means putting pieces of hardware into the rock that support your weight, instead of holding on with your hands and feet. You tie yourself to a rope, and you clip the rope to all the hardware you place, so that if the piece you're standing on pulls out, the ones below it will catch your fall.

Cryogenics was a perfect place to practice aid-climbing, I thought, because it was mostly overhanging. If you fell, you wouldn't slide painfully down the rock; instead, you would fall a short way and then dangle from the rope. Or you might fall until you hit the ground—but that possibility didn't seem too likely. The other good thing about Cryogenics, I thought, was that it had a crack a couple of fingers wide going most of the way up, so I could place as much gear as I wanted.

My Dad was agreeable, so I racked up and hopped on the route. Only then did I realize that my plan had some drawbacks. The rock wasn't great inside the crack. It took a lot of gear, but it was not easy for me to get a really bomb-proof piece in. And though it was a short pitch, it really ate up the gear, so that when I got near the top, I was running seriously short of the most effective pieces. The last bit was the hardest to free-climb, and I had almost no gear left. But I was almost there! I placed a marginal nut in a flaring crack. I stepped up on it, and it held. I placed a hex in the same crack. I stepped up on the hex, it pulled out, and I fell. What happened next passed in a flash, and I don't remember it, but it's easy enough to reconstruct.

The nut pulled out. I fell into empty space. The next nut pulled out. Climbers call this phenomenon "zippering," because it's like unzipping a zipper. If enough pieces pull out, then you hit the ground. Every time a piece pulls, the one below it has to withstand a harder pull, because you have gathered more speed and more momentum. With a jerk, I came up short on the next piece. It was a cam, which is the most sophisticated piece of hardware in a climber's arsenal. It was not ideally placed, but it held. My Dad, who was holding the rope while seated on the ground below, skidded forward as the rope between us came taut.

And that was it. I spent some time studying the cam that held my fall. It seemed to have pulled and turned a little, but it was still OK. I improved a couple of the placements of pieces just below it, then lowered off the climb. I walked around for a few minutes thinking about how hard the ground was. I went back up the rope, recovered most of my gear, and called it a day.

What can we learn from my experience on Cryogenics? Well, the first thing is this: when aid-climbing, you should stop when you run out of gear.

GRAVITY AND BLACK HOLES

The second thing is that falling is not a problem. Landing is the problem. I walked away without a scratch because I didn't hit the ground. (I did get a nosebleed a few minutes later.) Coming up short on that cam felt like a jerk, but it was a mild sort of jerk compared to the awful, crunching impact of hitting the ground.

There's a profound lesson about gravity that you can learn from falling. While you're doing it, you feel no gravity. You feel weightless. You get a similar feeling, to a lesser degree, when an elevator starts going down. I'd like to tell you that I have some sort of deeper appreciation for gravity based on my intimate personal experience with falling. The fact is, on Cryogenics, either I didn't have time to appreciate the experience, or I was too freaked out for rational thought to kick in.

Black holes

What would it be like to fall into a black hole? Would there be an awful, crunching impact? Or would you just fall forever? Let's take a quick tour of the properties of black holes to find out the answers.

First of all, a black hole is an object from which no light can escape. "Black" is meant to convey the total darkness of such an object. The surface of a black hole is called its horizon, because someone outside the horizon can't see what happens inside. That's because seeing involves light, and no light can get out of a black hole. Black holes are believed to exist at the centers of most galaxies. They are also thought to be the final stage of evolution of very massive stars.

The strangest thing of all about black holes is that they are just empty space, except for a "singularity" at their center. This might seem like nonsense: how can the most massive

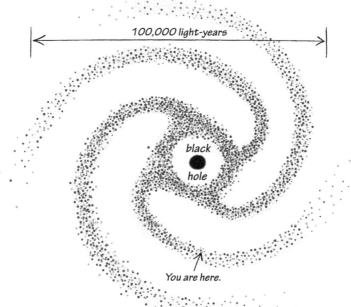

Our galaxy, the Milky Way, probably has a black hole at its center. The mass of the black hole is thought to be about four million times the mass of the sun. From Earth's perspective, it lies in the direction of the Sagittarius constellation. It's about 26,000 light-years away from us. The black hole is much smaller than the size depicted here, and so is the region around it empty of stars.

object in the galaxy be empty space? The answer is that all the mass inside the black hole collapses into the singularity. We actually don't understand what happens right at the singularity. What we do understand is that the singularity distorts spacetime in such a way that there is a horizon surrounding it. Anything that gets within the horizon will eventually be drawn into the singularity.

GRAVITY AND BLACK HOLES

Imagine a rock climber unfortunate enough to fall into a black hole. Crossing the horizon wouldn't cause him any injury, because there's nothing there: it's empty space. The rock climber probably wouldn't even notice when he fell through the horizon. The trouble is, nothing could then stop his fall. First of all, there's nothing to hold onto— remember, it's all empty space inside a black hole except for the singularity. The climber's only hope is his rope. But even if the rope were clipped to the most bomb-proof piece of gear in history, it wouldn't do any good. The gear might hold, but the rope would break, or it would stretch and stretch until the climber hit the singularity. When that happened, presumably there really would be an awful, crunching impact. But it's hard to know for sure, because no one could observe it except the climber himself. That's because no light can get out of a black hole!

The main thing to take away from this discussion is that gravity's pull inside a black hole is absolutely irresistible. Once he passed through the horizon, our unfortunate rock climber could no more stop his fall than he could stop time. And yet, nothing would "hurt" until he hit the singularity. Up until then, all he'd be doing is falling in empty space. He would feel weightless, as I must have felt while falling off of Cryogenics. This highlights a fundamental premise of general relativity: A freely falling observer feels like he's in empty space.

Here's another analogy that might help. Imagine a lake in the mountains, drained by a small, swift stream. Fish in the lake know not to venture too close to the mouth of that stream, because if they start down the stream, it's impossible for them to swim fast enough to escape the current and get back up into the lake. Unwise fish who let themselves be drawn into the stream don't get hurt (at least not right away),

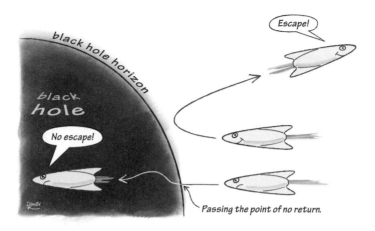

The black hole horizon is the point of no return. A spaceship can go
close to it and then turn around and escape. But if the spaceship goes
inside the horizon, it can never get back out.

but they have no choice but to keep on being drawn down-
stream. The lake is like spacetime outside a black hole, and
the interior of the black hole is like the stream. The singularity would be like some sharp rocks upon which the stream
breaks, and upon which any fish in the stream would come to
an immediate and sanguinary end. You might imagine other
possibilities: for example, maybe the stream leads to another
lake which a fish can reach in perfect safety and comfort.
Likewise, maybe there isn't a singularity inside a black hole
after all, but instead a tunnel to another universe. That seems
a bit far-fetched, but given that we don't understand singularities and can't find out what's inside a black hole except by
falling into one, it can't be ruled out completely.

In an astrophysical setting, there's an important caveat to
the idea that you feel nothing as you approach a black hole
and then cross the horizon. The caveat has to do with tidal

GRAVITY AND BLACK HOLES

forces. Tidal forces are so named because of how ocean tides arise. The moon pulls on the earth more strongly on the side closest to it. The sea rises on that side in response to the moon's pull. The sea also rises on the opposite side of the earth, which might seem pretty counterintuitive. But think of it this way: the middle of the earth is getting pulled toward the moon more than the oceans on the side opposite to the moon. Those oceans rise because they're getting left behind. Everything else moves toward the moon more than they do, because everything else is closer to the moon, and more affected by it.

When an object like a star gets close to a black hole, there's a similar effect. The parts of the star closer to the black hole are pulled in more strongly, and the star gets elongated as a result. As the star approaches the black hole horizon, it eventually gets shredded to bits. This shredding involves both tidal forces and the star's rotational motion around the black hole. In order to avoid unnecessary complications, let's ignore rotation and just think about a star falling straight into the black hole. Let's simplify further by replacing the star by two freely falling observers, initially separated by a distance equal to the diameter of the star. What I have in mind is that the trajectories of these two observers are supposed to be similar to the trajectories of the parts of the star nearest and farthest from the black hole. I'll refer to the observer who starts off closer to the black hole as the near-side observer. The other one is the far-side observer. The black hole pulls more strongly on the near-side observer, just because she's closer. So she starts falling faster than the far-side observer, with the result that the observers wind up getting farther apart as they fall. From their point of view, there appears to be a force pulling them apart. That apparent force is the tidal force, which is simply an expression of the fact that at any given time, gravity pulls harder on the near-side observer than on the far-side observer.

It may help to consider another analogy. Imagine a line of cars that start all bunched up on slow traffic. When the first car gets to a place where it can speed up, it pulls away from the car behind it. Even when the second car speeds up in the same place, it will remain a greater distance behind the first one. That's very similar to the way the distance between the near-side and far-side observers increases when they start falling toward the black hole. The elongation of a star falling toward a black hole is essentially the same phenomenon— except that in order to give a fully realistic description, one would have to account for the rotational motion of the star around the black hole, and eventually also the peculiar distortion of time near a black hole horizon.

Modern experiments aim to detect events like stars falling into black holes, or two black holes falling into one another. One of the key ideas is to detect the blast of gravitational radiation that occurs when the two massive objects are merging. Gravitational radiation is not something you can see with the naked eye, because what we see is light. Gravitational radiation is something completely different. It is a traveling wave of spacetime distortion. It carries energy, just like light does. It has a definite frequency, like light does. Light is composed of photons—little bits, or quanta, of light. And gravitational radiation, we think, is likewise composed of little quanta called gravitons. They obey the same relation $E = h\nu$ between energy E and frequency ν that photons do. They travel at the speed of light and are massless.

Gravitons interact with matter much more weakly than photons do, so there is no hope of detecting them through some analog of the photoelectric effect. Instead, the scheme for detection cuts directly to the fundamental nature of gravitational radiation. When a gravity wave passes between two objects, the distance between them fluctuates. That's because the spacetime between them itself fluctuates. The detection scheme, then, is to measure the distance between two objects very accurately and wait for it to fluctuate. If this scheme works, it will open up a whole new view of the universe. It will also be a spectacularly direct confirmation of the theory of relativity, which predicts gravitational radiation, whereas the previous Newtonian theory of gravity did not.

The general theory of relativity

I've actually already told you a lot about general relativity, indirectly. It's the theory of spacetime that describes black holes and gravitational radiation. In general relativity, spacetime is not a static stage on which events occur, but a dynamic, curved geometry. Gravitational waves are ripples in this geometry, propagating like the ripples you get from throwing a stone into a lake. A black hole is like a stream draining the lake. Both analogies are imperfect. The main missing ingredient is a new version of time dilation that goes to the heart of general relativity.

First, let me remind you about time dilation in special relativity. In special relativity, spacetime remains fixed. It's all about how objects behave when they move relative to one another. Time dilation describes how time slows down when you've moving. The faster you move, the more time slows down. When you reach the speed of light, time stops.

Here is the new feature of time dilation in general relativity. Time slows down when you're deep down in a gravitational well, like the one created by a massive star. When you get to a black hole horizon, time stops.

But wait! I've previously claimed that there's nothing special about a black hole horizon, except that once you fall in, you can't get out again. Crossing a horizon isn't a special experience. How can this be if time stops at a black hole horizon? The resolution is that time is a matter of perspective. A rock climber falling through a horizon experiences time differently from the way you would if you hovered just a little bit above the horizon. An observer far from the black hole has yet a different experience of time. From the point of view of a distant observer, it takes infinitely long for anything to fall into a horizon. If such an observer watched a rock climber fall into the black hole, it would seem like the climber crept ever closer to the horizon but never quite fell in. According to the climber's own sense of time, it takes only a finite time for him to fall in, and only a finite additional time to get to the center of the black hole where the singularity lurks. I would say that time slows down for the climber, because a second for him corresponds to a much longer time for the distant observer. Time also slows down for an observer who hovers slightly above the black hole. The closer he is to the horizon, the more time slows down.

All this seems terribly abstract, but it has real-world consequences. Time runs slower at the surface of the Earth than it does in outer space. The difference is slight: it's a little less than one part in a billion. But it matters for the Global Positioning System (GPS). The reason is that precision time measurements are part of what makes it possible for GPS to locate objects precisely on the surface of the Earth. Those time measurements suffer time dilation effects, both because the satellites are moving and because they're not as far down in Earth's gravitational well as we are. Accounting properly for time dilation effects is a crucial ingredient in making GPS work as well as it does.

I remarked earlier that there's a connection between time dilation and kinetic energy. Let me remind you. Kinetic energy is the energy of motion. Time dilation occurs when you are in motion. If you move so fast that you double your rest energy, then time runs half as fast. If you move so fast that you quadruple your rest energy, then time runs a quarter as fast.

There's something very similar in the case of gravitational redshift, but it relates to gravitational energy. Gravitational energy is the amount of energy that you can gain by falling. If a piece of space debris falls to the Earth, the energy it gains by falling is a little less than a billionth of its rest mass. It's no accident that this is the same tiny fraction that characterizes how much gravitational redshift there is on the surface of the Earth. Time running at different rates in different places *is* gravity. In fact, that's *all* gravity is, provided gravitational fields are not too strong. Things fall from places where time runs faster to where it runs slower. That downward pull you feel, and which we call gravity, is just the differential rate of time between high places and low places.

Black holes aren't really black

String theorists' interest in black holes comes in large part from their quantum mechanical properties. Quantum mechanics turns the defining property of black holes on its head. No longer are black hole horizons black. They glow like a live coal. But their glow is very faint, very cold—at least, if we're talking about astrophysical black holes. The glow of a black hole horizon means that it has a temperature. That temperature is related to how strong gravity is at the surface of the black hole. The larger a black hole is, the lower its temperature—at least, if we're talking about astrophysical black holes.

Temperature is going to come up again, so we'd better discuss it more carefully. The right way to understand it is in terms of thermal energy, or heat. The heat in a mug of tea comes from the microscopic motion of the water molecules. When you cool water, you're sucking out thermal energy. Each water molecule moves less and less vigorously. Eventually the water freezes into ice. That happens at zero degrees Celsius. But the water molecules in the ice are still moving a little: they vibrate around their equilibrium positions in the ice crystal. As you cool ice down further and further, these vibrations get weaker and weaker. Finally, at −273.15 degrees Celsius, all such vibrations stop—almost: the water molecules are as fixed in their equilibrium positions as quantum uncertainty allows. You can't make something colder than −273.15 degrees Celsius (which is −459.67 degrees Fahrenheit) because there's no thermal energy left to suck out of it. This coldest of cold temperatures is called absolute zero temperature.

It's important that quantum mechanics prevents water molecules from ceasing to vibrate altogether, even at absolute zero temperature. Let's explore this a little more. The uncertainty relation is $\Delta p \times \Delta x \geq h/4\pi$. In an ice crystal, you know fairly precisely where each water molecule is. That means Δx is fairly small: certainly less than the distance between neighboring water molecules. If Δx is fairly small, it means that Δp cannot be too small. So, according to quantum mechanics, the individual water molecules are still rattling around a little, even when they're frozen solid in a cube of ice at absolute zero. There is some energy associated with this motion, which goes by the name "quantum zero-point energy." We've actually encountered it before when discussing the hydrogen atom. If you recall, I compared the lowest energy of an electron in a hydrogen atom to the fundamental frequency of a piano string. The electron is still moving. Both its position and its momentum have some uncertainty. Sometimes people describe this by saying that the electron undergoes quantum fluctuations. Its ground state energy can be termed the quantum zero-point energy.

In summary, there are two types of vibrations happening in an ice cube: thermal vibrations and quantum fluctuations. You can get rid of the thermal vibrations by cooling the ice down to absolute zero. But you can't get rid of the quantum fluctuations.

The idea of absolute zero temperature is so useful that physicists often quote temperatures in reference to it. This way of quoting temperature is called the Kelvin scale. One degree Kelvin—or, more commonly, one Kelvin—is one degree above absolute zero, or −272.15 degrees Celsius. 273.15 Kelvin is 0 degrees Celsius, the temperature where ice melts. If you measure temperature on the Kelvin scale, then the typical energy of thermal vibrations is given by a simple equation: $E = k_B T$, where k_B is called Boltzmann's constant. For example, at the melting point of ice, this formula says that the typical energy of the thermal vibrations of a single water molecule is a fortieth of an electron volt. This is almost a hundred times smaller than the amount of energy it takes to knock an electron out of sodium, which, as you may recall from chapter 2, is 2.3 electron volts.

Here are a few more interesting temperatures, just to get a feel for the Kelvin scale. Air turns into a liquid at about 77 Kelvin, which is −321 degrees Fahrenheit. Room temperature (say, 72 degrees Fahrenheit) is about 295 Kelvin.

CHAPTER THREE

Physicists are able to cool small objects down to less than a thousandth of a Kelvin. On the other extreme, the surface of the sun is a little less than 6000 Kelvin, and the center of the sun is about 16 million Kelvin.

Now, what does all this have to do with black holes? A black hole doesn't seem to be made up of little molecules whose vibrations can be classified as thermal or quantum. Instead, a black hole is made up only of empty space, a horizon, and a singularity. It turns out that empty space is quite a complicated thing. It experiences quantum fluctuations that can be roughly described as spontaneous creation and destruction of pairs of particles. If a pair of particles is created near a black hole horizon, then it can happen that one of the particles falls into the black hole and the other one escapes, carrying energy away from the black hole. This kind of process is what gives a black hole a non-zero temperature. To put it succinctly, a horizon converts some of the omnipresent quantum fluctuations of spacetime into thermal energy.

Thermal radiation from a black hole is very faint, corresponding to a very low temperature. Consider, for example, a black hole formed in the gravitational collapse of a heavy star. It might contain a few times as much mass as the sun. Its temperature would then be about twenty billionths of a Kelvin, or 2×10^{-8} Kelvin. The black holes at the centers of most galaxies are much heavier: millions or even a billion times heavier than the sun. The temperature of a black hole five million times heavier than the sun would be about a hundredth of a trillionth of a Kelvin: that is, 10^{-14} Kelvin.

What fascinates string theorists is not so much the extreme lowness of the temperature of black hole horizons, but the possibility of describing certain objects in string theory, known as D-branes, as very small black holes. These very small black holes can have a wide range of temperatures, from absolute zero to arbitrarily high values. String theory relates the temperature of small black holes to thermal vibrations on the D-branes. I'll introduce D-branes more carefully in the next chapter, and I'll tell you more about how they relate to small black holes in chapter 5. This relation is at the heart of recent efforts to understand what happens in heavy ion collisions using string theory, as I'll discuss in chapter 8.

CHAPTER THREE

Chapter F O U R

STRING THEORY

WHEN I WAS A SOPHOMORE AT PRINCETON, I TOOK A COURSE on Roman history. It was mostly about the Roman Republic. It's fascinating how the Romans combined peaceful and military achievements. They evolved an unwritten constitution and some degree of representative democracy while simultaneously overpowering first their neighbors, then the Italian peninsula, and finally the whole of the Mediterranean and beyond. Equally fascinating is how the civil strife of the late republic ended in the tyranny of the empire.

Our language and legal system are filled with echoes of ancient Rome. For an example, look no further than the back of a quarter. If it was coined before 1999, then it shows an eagle perched on a bundle of sticks. This bundle is a fasces, a Roman symbol of strength and authority. The Romans also made influential contributions to literature, art, urban architecture and planning, and military tactics and strategy. The eventual adoption of Christianity in the Roman Empire helps account for Christianity's prominence today.

As much as I enjoy Roman history, I wouldn't bring it up if it didn't remind me of what I really want to talk about: string theory. We are deeply influenced by the Romans, but we are separated from them by a span of many centuries. String theory, if correct, describes physics at an energy scale far higher than we are able to probe directly. If we could probe the energy scales that string theory describes directly, then we would presumably see the various exotic things I am going to tell you about: extra dimensions, D-branes, dualities, and so forth. This exotic physics underlies the world we experience (assuming string theory is correct), just as Roman civilization underlies our own. But string theory is separated from our experience of the world—not by centuries of time, but by a similarly vast gulf in energy scales. Particle accelerators would have to be a hundred trillion times more powerful than the ones going into operation today to reach the range of energies where we think extra dimensions open up and stringy effects could be observed directly.

This gulf in energy scales brings us to the most uncomfortable aspect of string theory: it's hard to test. In chapters 7 and 8 I'll tell you about efforts to link string theory to experiment. In this chapter and the next two, I will instead try to convey string theory on its own terms, without reaching for connections to the real world except as explanatory devices. Think of these chapters as analogous to a brief summary of Roman history. The narrative of Rome has many twists and turns. It's sometimes hard to follow. But we study the Romans to understand not only their world, but through it our own. String theory also has some surprising twists and turns, and I expect my explanations of them may not always be easy to get through. But there is at least a good chance

CHAPTER FOUR

that a deep understanding of string theory will eventually be the basis of our understanding of the world.

In this chapter, we'll take three important steps toward understanding string theory. The first is to understand how string theory resolves a fundamental tension between gravity and quantum mechanics. The second is to understand how strings vibrate and move in spacetime. The third is to get a glimpse of how spacetime itself emerges from the most widely used mathematical description of strings.

Gravity versus quantum mechanics

Quantum mechanics and general relativity are the great triumphs of early twentieth-century physics. But they turn out to be hard to reconcile with one another. The difficulty hinges on a concept called renormalizability. I will describe renormalizability by comparing photons and gravitons, both of which we've discussed in previous chapters. The upshot is going to be that photons lead to a renormalizable theory (which means, a good theory), whereas gravitons lead to a non-renormalizable theory—which is really no theory at all.

Photons respond to electric charge, but they are not themselves charged. For example, an electron in a hydrogen atom is charged, and when it jumps from one energy level to another, it emits a photon. That's what I mean by photons responding to charge. Saying that photons are not themselves charged is the same as saying that light doesn't conduct electricity. If it did, you'd get a shock from touching something that had been in the sun too long. Photons do not respond to one another because they only respond to electric charge.

Gravitons respond not to charge, but to mass and energy. Because they carry energy, they respond to themselves. They self-gravitate. This might not seem too problematic, but here's how you get into trouble. Quantum mechanics teaches us that gravitons are particles as well as waves. A particle, by hypothesis, is a pointlike object. A pointlike graviton gravitates more strongly the closer you get to it. Its gravitational field can be understood as the emission of other gravitons. To keep track of all these gravitons, let's call the original one the mother graviton. We'll refer to the gravitons it emits as daughter gravitons. The gravitational field very close to the mother graviton is very strong. That means that its daughter gravitons have enormous energy and momentum. This follows from the uncertainty relation: daughter gravitons are observed within a very small distance Δx of the mother graviton, and so their momentum is uncertain by a large amount Δp, such that $\Delta p \times \Delta x \geq h/4\pi$. The trouble is that gravitons also respond to momentum. The daughter gravitons will themselves emit gravitons. The whole process runs away: you can't keep track of the effects of all these gravitons.

Actually, there's something similar that happens near an electron. If you probe the electric field very close to it, you provoke the electron into emitting photons with a large momentum. That seems innocuous, because we've learned that photons can't emit other photons. The trouble is, they can split into electrons and anti-electrons, which then emit more photons. What a mess! The amazing thing is that, in the case of electrons and photons, you actually can keep track of this multiplicity of particles cascading from one another. One speaks of a "dressed" electron to describe the electron and its cloud of progeny. The technical term for its progeny is virtual particles. Renormalization is the mathematical method

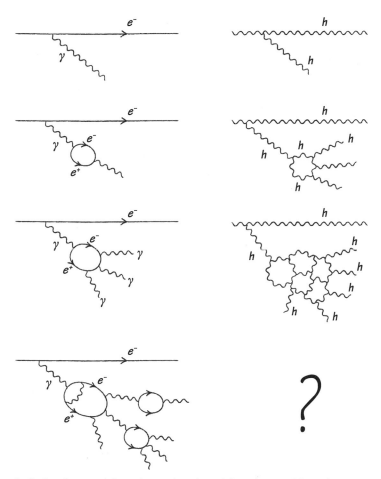

Left: An electron (e^-) produces virtual particles: photons (γ), positrons (e^+), and more electrons. The cascade of particles is slow enough to keep track of mathematically, using renormalization. Right: A graviton (h) produces so many virtual gravitons so prolifically that renormalization can't keep track of them.

for keeping track of them all. The spirit of renormalization is that an electron all by itself may have infinite charge and infinite mass, but once it is dressed, its charge and mass become finite.

The trouble with gravitons is that you can't renormalize the cloud of virtual gravitons that surround them. General relativity—the theory of gravity—is said to be non-renormalizable. This might seem like an arcane technical problem. There's a faint chance that we're just looking at the problem the wrong way. There's also a chance, perhaps less faint, that a relative of general relativity, called maximal supergravity, is renormalizable. But I, along with most string theorists, feel pretty sure that there is a fundamental difficulty merging quantum mechanics and gravity.

Enter string theory. The starting assumption is that particles are not pointlike. Instead they are vibrational modes of strings. A string is infinitely thin, but it has some length. That length is small: about 10^{-34} meters, according to conventional ideas about string theory. Now, strings respond to one another in a fashion similar to gravitons. So you might worry that the whole problem with clouds of virtual particles—actually, virtual strings—would get out of control, just as it did for gravitons. What stops this problem from arising is that strings aren't pointlike. The whole difficulty with gravity arose because point particles are, by assumption, infinitely small—hence the term "point particle." Replacing gravitons with vibrating strings smoothes out the way they interact with one another. One way to say it is that when a graviton splits into two, you can identify an instant of time and a position in space where the split occurred. But when a string splits, it looks like a pipe branching. At the branch point, no part of the pipe wall is breached: the Y is a smooth,

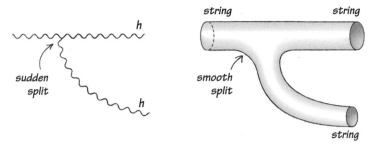

A graviton splits suddenly. A string splitting occurs over some region of spacetime, so it's more gentle.

solid piece of pipe, just in an unusual shape. What all this is leading up to is that the splitting of a string is a more gentle event than the splitting of a particle. Physicists say that string interactions are intrinsically "soft," whereas particle interactions are intrinsically "hard." It is this softness that makes string theory more well-behaved than general relativity, and more amenable to a quantum mechanical treatment.

Strings in spacetime

Let me remind you briefly of our earlier discussion of a vibrating piano string. When you stretch it tightly between two pegs and pluck it, it vibrates with a definite frequency. The frequency is the number of vibrations per second. A piano string also has overtones of vibration: higher pitches that blend with its fundamental frequency to produce the particular sound you associate with a piano. I drew an analogy to the behavior of an electron in a hydrogen atom: it too has a preferred vibrational mode, corresponding to its lowest energy level, as well as other vibrational modes, corresponding to higher energy levels. This analogy might have left you a little cold: what does an electron in a hydrogen atom really have to do with standing waves on a stretched string? It's more like a particle rotating around the atomic nucleus—like an infinitesimally small planetoid rotating around a tiny little sun. Right? Well, yes and no: quantum mechanics says the particle picture and the wave picture are deeply related, and the quantum mechanical motion of the electron around the proton actually can be described as a standing wave.

We can be much more direct in comparing a piano string with the strings of string theory. To distinguish between different types of strings, let me call the strings of string theory "relativistic strings." This term anticipates deep insights that we'll discuss soon, namely that strings incorporate the theories of relativity, both special and general. For the moment, I want to talk about a string theory construction that is as close as I can manage to a stretched piano string. Relativistic strings are allowed to end on objects called D-branes. If we suppress the effects of string interactions, D-branes are infinitely heavy. We'll discuss them in a lot more detail in the next chapter, but for now they're just a crutch for our understanding. The simplest of D-branes is called a D0-brane, usually pronounced as "dee-zero brane." It's a point particle. You might feel bothered by the fact that point particles just appeared again in the discussion. Wasn't string theory supposed to get rid of them? The fact is, it did for a while, and then in the middle 1990s point particles came back, along with a whole zoo of other things. But I'm getting ahead of the story. What I want is a string theory analog of the tuning pegs in a piano, and D0-branes are so appropriate that I can't resist introducing them. So, let's stretch a relativistic string between two D0-branes, like we'd stretch a piano string between two pegs. The D0-branes aren't attached to anything, but they don't move because they are infinitely heavy. Pretty weird stuff, eh? I'll say more about D0-branes in the next chapter. What I really want to talk about here is the stretched string.

The lowest energy level of the stretched string has no vibrations. Well, almost none. There's always a little quantum mechanical vibration, and that's going to be important in a minute. The right way to understand the ground state is that it has as little vibrational energy as quantum mechanics allows. The relativistic string has excited states where it is vibrating, either at its fundamental frequency, or in one of its overtone frequencies. It can vibrate simultaneously in several different frequencies, just like a piano string does. But just as the electron in a hydrogen atom cannot move in an arbitrary fashion, so too a relativistic string cannot vibrate arbitrarily. The electron has to choose among a series of energy levels with finite spacing from one another. Similarly, the string has to choose among a series of vibrational states. The vibrational states have different energies. But energy and mass are related through $E = mc^2$. So the different vibrational states of a string have different masses.

It would be nice if I could tell you that the vibrational frequencies of a string correspond in a simple way to its energy, just as the equation $E = h\nu$ connects the frequency and energy of a photon. There's something like that going on, but unfortunately it's not quite so simple. The total mass of a string comes from several different contributions. First, there's the rest mass of the string: the mass it has on account of being stretched from one D0-brane to the other. Next, there's the vibrational energy in each overtone. This contributes to the mass, because energy is mass according to $E = mc^2$. Finally, there's a contribution coming from the minimal amount of vibration allowed by quantum uncertainty. This contribution

58

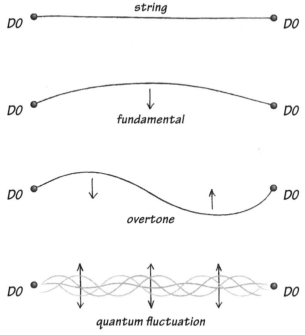

Motions of a string stretched between two D0-branes.

from quantum fluctuations is called zero-point energy. The term "zero-point" is supposed to remind us that this quantum contribution can't be gotten rid of. The contribution of the zero-point energy to the mass is *negative*. That's strange. Really strange. To understand just how strange, consider the following. If you restrict attention to a single vibrational mode of the string, the zero-point energy is positive. Higher overtones make bigger positive contributions to the zero-point energy. But when you sum them all up in an appropriate way, you get a negative number. If that isn't bad enough, here's worse news: I lied a little in saying that the contribution of zero-

CHAPTER FOUR

point energy to the mass is negative. All these effects—rest
mass, vibrational energies, and zero-point energy—add up to
the total mass squared. So if the zero-point energy dominates,
the mass squared is negative. That means the mass is *imaginary*,
like $\sqrt{-1}$.

Before you dismiss all this as nonsense, let me hasten to add
that a large swath of string theory is devoted to getting rid of
the awful problem I described in the previous paragraph. To
put this problem in a nutshell, a relativistic string in its least
energetic quantum state has negative mass squared. A string
in such a state is called a tachyon. Yes, those are the same
tachyons that Star Trek characters confront in about every
other episode. They're obviously bad news. In the setup I described, where the string is stretched between two D0-branes,
you can get rid of them just by separating the D0-branes far
enough so that the contribution to the mass from stretching
the string is bigger than the quantum fluctuations. But when
there aren't any D0-branes around, there are still strings. Instead of ending on something, they close upon themselves.
They aren't stretched out at all. They can still vibrate, but they
don't have to. The only thing they can't avoid doing is fluctuating quantum mechanically. And as before, those zero-point
quantum fluctuations tend to make them tachyonic. This is
bad, bad news for string theory. The modern view is that
tachyons are an instability, similar to the instability of a pencil
balanced on its point. If you're extremely persistent and skill-
ful, maybe you can balance a pencil that way. But the least
breath of wind will knock it over. String theory with tachyons is kind of like a theory of the motion of a million pencils,
distributed throughout space, all balanced on their points.

Let me not paint things too black. There is a saving grace
of tachyons. Let's accept that the ground state of a string is

a tachyon, with negative mass squared: $m^2 < 0$. Vibrational energy makes m^2 less negative. In fact, if you play your cards right, the smallest increment of vibrational energy that quantum mechanics allows makes m exactly 0. That's great, because we know there are massless particles in nature: photons and gravitons. So if strings are to describe the world, there must be massless strings—more precisely, there must be vibrational quantum states of strings that are massless.

I said something about playing your cards right. Exactly what does that mean? Well, it means that you need 26 dimensions of spacetime. You probably knew this was coming, so I won't apologize for it. There are several arguments for 26 dimensions, but most of them are pretty mathematical, and I'm afraid I couldn't make them sound at all convincing. The argument I have in mind hinges on the following points. You know you want massless string states. You know there are zero-point quantum fluctuations that push m^2 negative. And you know there are vibrational modes that push m^2 in the other direction. The smallest amount of vibrational energy doesn't depend on the dimension of spacetime. But the zero-point quantum fluctuations do. Think of it this way: when something vibrates, like a piano string, it does so in a definite direction. A piano string vibrates in the direction in which it was struck. In a grand piano, that's up and down, not side to side. Vibrations pick out a direction and ignore all others. In contrast, zero-point quantum fluctuations go in every possible direction. Every new dimension you introduce gives the quantum fluctuations another direction to explore. More directions means more zero-point fluctuations, so a more negative contribution to m^2. All that's left is to ask how things balance out between vibrations and the irreducible zero-point quantum fluctuations. It's a matter of calculation.

CHAPTER FOUR

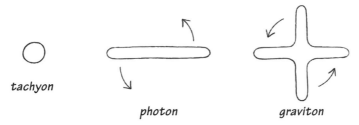

tachyon

photon   graviton

Cartoons of the quantum states of a string that make it act like a
tachyon, or a photon, or a graviton.

It turns out that the minimal amount of vibration cancels out
against 26 dimensions' worth of quantum fluctuation, leading to massless string states as desired. Look on the bright
side. It could have been 26 and a half dimensions.

If you're getting confused between vibrations of a string and
the zero-point quantum fluctuations, don't be disappointed.
They're really similar. The only difference is that vibrations
are optional, and zero-point quantum fluctuations are not.
Zero-point fluctuations are the minimum amount of vibration required by uncertainty. Extra vibrations on top of that
can still be quantum mechanical. It helps me to think of the
vibrations as giving a string a characteristic shape: maybe
circular, maybe cloverleaf, maybe folded over and spinning.
Those different shapes are supposed to correspond to different particles. But speaking of the shape of a vibrating string
is imprecise, because all the vibrations are quantum mechanical. Better is to say that the different quantum modes
of vibration of a string correspond to different particles. The
shapes are mental pictures that help us visualize some of the
properties of those quantum vibrations.

To summarize, we've got good news, bad news, and
worse news. Strings have vibrational modes and can act like

a photon or a graviton. That's the good news. They can only do this in 26 dimensions. That's the bad news. There's also a vibrational mode of a string that has imaginary mass, the tachyon. It signals an instability of the whole theory. It doesn't get much worse than that.

Superstring theory cures the tachyon problem, and it lowers the number of dimensions from 26 to 10. It also produces new vibrational modes that allow strings to act like electrons. Pretty cool stuff, all in all. If only there were a super-duper-string theory that cut the number of dimensions down to 4, maybe we'd be in business. I say this only half in jest. There actually is a version of super-duper-string theory, the more technical name for which is string theory with extended local supersymmetry. It cuts the number of dimensions down to 4. Unfortunately, those dimensions come in pairs, so you either get four spatial dimensions and no time, or two spatial dimensions and two times. Not good. We need three dimensions of space and one time. Of the ten dimensions that superstring theory requires, nine are spatial dimensions and one is time. To relate superstring theory to the world, somehow we have to do away with six of the nine spatial dimensions.

There's a lot I would like to tell you about superstrings, but most of it has to wait until later chapters. Let me focus here on a synopsis of how the tachyon problem gets cured. The superstring fluctuates not just in space and time but in other, more abstract ways. These other types of fluctuations go partway toward solving the tachyon problem, but not all the way. There's still a vibrational mode with negative mass squared. The key to the story is that if you start with the vibrational modes representing photons, electrons, and other particles that we want, no matter how you collide them, you can never make a tachyon. It feels like the whole theory is

CHAPTER FOUR

still balanced on a knife edge. But it has a special symmetry that helps it stay balanced. That symmetry is called super-symmetry. Physicists hope to find evidence for supersymmetry in the next few years. If it is found, many of us would take it as confirmation of superstring theory. I'll discuss this more in chapter 7.

Spacetime from strings

I've talked a lot about strings vibrating and fluctuating in spacetime. Let's take a step back and ask, just what is space? Just what is time? One view is that space derives meaning only through the objects present in space. What space describes is the distance between objects. A similar view of time is that it's meaningless by itself, but only describes the sequence of events. To make this more definite, consider a pair of particles, A and B. The conventional view is that each of them moves on some trajectory through spacetime, and they collide if the trajectories cross. Perhaps there's nothing wrong with that. But let's try to take the alternative view that space and time have no meaning in the absence of the particles. What would that mean? Well, to describe the trajectory of particle A, we could specify its position as a function of time. And the same for particle B. If we could do that, then perhaps we could disregard space and time except as represented by the evolving position of the particles. We would still know if the particles collided, because they would have the same position and the same time when they hit.

If this seems too abstract, think of the particles as racecars equipped with GPS devices and clocks. Let's suppose the GPS devices record where the racecars are once a second. What could we learn from examining the records from the GPS devices? Well, let's suppose the racecars all move on the same racetrack. From the GPS records, the first thing you would notice is that the cars keep coming back to the same place after traveling a fixed distance—the distance around the track. So you would say, Aha! The cars travel on a circular track. Next, let's say you notice that the cars speed up and slow down a lot. After scratching your head a bit, you might figure out that the racetrack isn't circular after all! Instead it has curves, where the racecars have to slow down, and straightaways, where they can go fast. You might also observe that all the cars for which you have records go around the track in the same direction. You could correctly conclude that there's a rule at the racetrack that everyone has to go the same way. Finally, you would notice that the cars have many near misses but seldom crash into one another. And you might reasonably deduce that a goal of car races is not to have collisions.

The upshot is that just by looking at the GPS records of a number of racecars, and doing a lot of detective work, you could figure out quite a bit about the racetrack and the rules of driving on it. It may seem that this is a bass-ackwards way of finding things out that you could much more easily discover by watching an actual race. But in truth, watching a race is a very complicated activity. You're in a stand off to the side of the track—which already means that the racetrack can't be all there is to spacetime. Watching means that photons are bouncing off the cars and going into your eyes, and that involves a lot of physics. It's really a lot simpler to say that the GPS records of where all the cars were, second by second, contains the essential information about what happened on the racetrack. With such records in hand, you don't have to ask about complicated things like observers in the stands and photons going hither and thither. You don't

CHAPTER FOUR

have to ask—in fact, you can't meaningfully ask—whether there is anything in the world beyond the racetrack. You don't even have to assume that the racetrack exists. Instead, you can deduce its existence and some of its properties by studying the records of how the cars moved.

A lot of string theory is done in a similar way. From the way strings move and interact, you deduce properties of space and time. This approach is called worldsheet string theory. The worldsheet is a way of recording how a string moves. It is like the second-by-second GPS record of where a car is on a racetrack. It's more complicated, though, for two reasons. First, a string can be long and floppy, so to say where it is, you have to say where every bit of it is. Second, as we've reviewed, strings usually live in 26 dimensions, or at least ten. These dimensions might be curved or rolled up in some complicated way. It is usually not possible to "stand to the side" and "look" at the geometry of spacetime the way you would watch cars at a racetrack. Meaningful questions are ones that can be phrased in terms of how strings move and interact. Spacetime itself, in the worldsheet approach, is only what the strings experience, rather than a fixed stage.

The string worldsheet is a surface. If you cut across it, you get a curve, and that curve is supposed to be the string. Cutting across the worldsheet in different ways is like checking the GPS record of a car at different moments. In order to say how the string moves in spacetime, you have to specify a point in space and a moment in time for each point on the worldsheet. Think of it as attaching a whole bunch of labels to the worldsheet. When you cut across the worldsheet, the curve you get still has those labels, so it "knows" what shape in space it's supposed to assume. The worldsheet as a whole is the surface that the string sweeps out in spacetime as it moves.

66

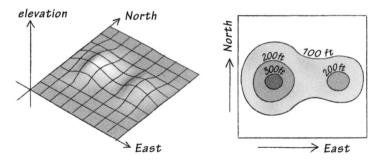

Left: Two hills separated by a saddle. Right: a topographical map of the
hills, with contours of constant elevation labeled.

You can appreciate what I mean by labeling the worldsheet by thinking about a topographical map (topo for
short). On a topo, there are lines of elevation, and each one
is labeled—or, if that's too many labels to read, sometimes
people label one line in five. Now, the topo itself is a perfectly flat piece of paper. But it represents terrain that can be
quite hilly.

One way to think of the string worldsheet is that it's like
a topographical map of how the string is supposed to move
in spacetime. But another point of view is that the string
worldsheet is all there is, and spacetime is no more than the
collection of labels you put on the worldsheet. In ordinary
topographical maps, the labels are elevations, so the collection of labels is just the range of possible elevations on the
surface of the Earth: about −400 meters to 8800 meters if
you exclude ocean floors. In worldsheet string theory, each
label specifies a location in 26 dimensions (or ten in the case
of superstrings). Some of those 26 dimensions can wrap back
around and reconnect to themselves, like a racetrack does.
The point is that the concept of spacetime emerges from

CHAPTER FOUR

how you label the worldsheet, just as you could say that elevation "emerges" from the way you label topographic maps.

Let's recap and then get to one of the main punch lines of worldsheet string theory. Usually we think of strings vibrating in spacetime. But space and time don't have to be absolute notions. It's better if they're not, because then some outside dynamical principle can control the shape of spacetime. It happens that way in string theory. In the worldsheet approach to string theory, spacetime is just the list of labels allowed in a description of how the string moves. In a quantum mechanical treatment, these labels fluctuate a little. Now, here is the real punch line. It turns out that you can keep track of these quantum fluctuations only if the spacetime obeys the equations of general relativity. General relativity is the modern theory of gravity. So quantum mechanics plus worldsheet string theory implies gravity. Pretty cool.

Explaining how you "keep track" of the quantum fluctuations of spacetime labels on the string worldsheet would take us into excessively technical territory. But there is a point of contact with the racecar analogy that may help your intuition. If you recall, I suggested that you might guess that the racetrack has straightaways and curves by noticing that the racecars slow down to traverse certain parts of the track and then speed up in other parts. Well, one thing a racetrack almost never has is corners where you have to turn really suddenly. That's because the cars would all have to stop at the corners, which would be no fun, and contrary to the spirit of a car race. Similarly, one of the things that the equations of general relativity almost entirely forbid is sharp corners in spacetime—usually called singularities. I say "almost" because singularities are allowed behind black hole horizons. For the most part, you can understand the absence of singularities in spacetime as analogous to the absence of corners on a racetrack. Strings can no more pass through most singularities than racecars can zip around a corner without stopping. But there are some exceptions. A fascinating and large subject in string theory is understanding the types of singularities that are allowed. Usually these singularities cannot be understood in general relativity. So string theory actually allows a somewhat richer class of spacetime geometries than relativity does. It turns out that the extra geometries string theory allows are in some cases related to branes, which we will encounter in the next chapter.

CHAPTER FOUR

Chapter F I V E

BRANES

In 1989, after my junior year in high school, I went to a physics camp. One of the things we did was to hear a lecture on string theory. About halfway through, one of the other students asked a sharp question. He said (more or less), "Why stop with strings? Why not work with sheets, or membranes, or solid three-dimensional chunks of quantum stuff?" The lecturer basically replied that strings seemed to be both difficult enough and powerful enough already, and that they seemed to be special in ways that membranes and solid chunks were not.

Fast forward about six years, to 1995: The whole string theory community was electrified by the advent of D-branes. D-branes are exactly what the sharp student had asked for in 1989. They are objects in string theory that can have any number of dimensions. This chapter is mostly about D-branes and some of their amazing properties. I'll start with a brief account of the second superstring revolution, which was a tide of new ideas that swept through the field in the mid-1990s. I'll tell you more precisely what a D-brane is, and I'll discuss the concept of symmetry and how it relates to D-branes. Next I'll describe how D-branes relate to black holes. At the end, I'll get to some discussion of M-theory, which is an eleven-dimensional theory that is necessitated by string theory but not entirely part of it.

The second superstring revolution

The perspective on string theory that I presented in the previous chapter is about what string theorists understood in 1989. They understood the danger of tachyons, the miraculous properties of the superstring, and the relationship between strings and spacetime. Another thing they understood, which I have scarcely mentioned, is compactification: the process of rolling up the six extra dimensions of superstring theory so as to be left with three dimensions of space and one of time. It all looked pretty good, because you had all the main ingredients of fundamental physics. Gravity was there. Photons were there. Electrons and other particles were there. Interactions among them were about what one wanted. Clever compactifications seemed to give just about the correct list of particles—a list that stretches well beyond the ones I've mentioned so far. But string theorists couldn't "close the deal" by producing a compactification that was just right, leading to precisely the physics that we observe in the real world.

Looking back on that period, there was another problem. It was strings, strings, strings, all day every day. The understanding of the string worldsheet was profound, but the very depth of that understanding may have temporarily blinded people to the possibilities that eventually got explored in the second superstring revolution. It's hard for me to trace the history of that period with complete accuracy, because I entered the field a little after the second revolution started. But it's clear that hints started accumulating that strings weren't the whole story. Before starting a detailed discussion of branes, it seems to me worthwhile to summarize some of those hints and to give an overview of what the second superstring revolution was about.

One hint was that interactions among strings became less and less controllable the more splitting and joining events there were. It was suggested that some new objects had to be added in order to handle string theory when splitting and joining interactions became strong. Another hint came from theories of supergravity. Supergravity is a low-energy limit of superstring theory. What I mean by a "low-energy limit" is that you throw away all but the lowest energy vibrational modes of the superstring. What you have left is the graviton and some other particles whose interactions are very precisely understood as long as they're not too energetic. It was observed that supergravity theories had some remarkable symmetries that weren't visible in the worldsheet description of string theory. That seemed to indicate that the worldsheet description was incomplete. The broadest hint came from the construction of branes. A brane is like a string, but it can have any number of dimensions of spatial extent. A string is a 1-brane. A point particle is a 0-brane. A membrane, which at any given instant of time is a surface, is a 2-brane. And there are 3-branes, 4-branes, 5-branes (two kinds!), 6-branes, 7-branes, 8-branes, and 9-branes. With so many different branes present in string theory, it started to seem implausible that everything could be understood in terms of strings alone. A final hint came from eleven-dimensional supergravity. It is a theory constructed using only two ideas: supersymmetry and general relativity. It has some connections to

72

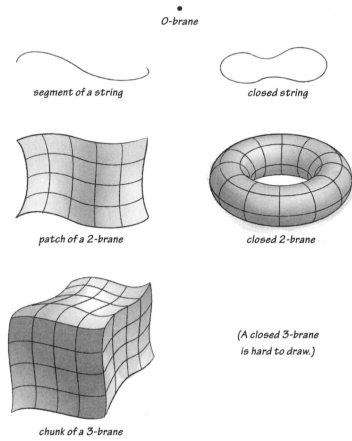

•
0-brane

segment of a string

closed string

patch of a 2-brane

closed 2-brane

*(A closed 3-brane
is hard to draw.)*

chunk of a 3-brane

0-branes, strings, 2-branes, and 3-branes. A string can close on itself to form a closed loop. A 2-brane can close on itself to form a surface without boundary. A 3-brane can do something similar, but it's hard to draw.

the supergravity theories that come out of string theory, and those connections were understood well before the second superstring revolution. But it wasn't at all clear how or whether it was related to worldsheet string theory. Worst of all, it didn't incorporate quantum mechanics, and therefore

CHAPTER FIVE

was viewed with skepticism by string theorists who were used to the idea that quantum mechanics and gravity are tightly intertwined. Eleven-dimensional supergravity was, in short, a mystery to string theorists: something close to what they were most interested in, but that didn't entirely make sense.

The field changed dramatically in a few short years in the middle 1990s, as these hints suddenly fell into a coherent pattern. Strings were still recognized as important, but it emerged that branes of various dimensions were also essential. At least in some circumstances, branes had to be placed on an equal footing with strings. In other circumstances, branes could be described as zero-temperature black holes. Eleven-dimensional supergravity also fit beautifully into the new circle of ideas. It seemed so central, in fact, that it got a new name: M-theory. More properly, M-theory is whatever consistent quantum theory has eleven-dimensional supergravity as its low-energy limit. Sadly, the second superstring revolution fell short of giving a full description of what M-theory really is. What became clear, however, is that with the new toolbox that branes provide, one could understand string theory in a new way. Especially surprising was the realization that when string interactions are very strong, new objects (often branes) offer simpler descriptions of the dynamics.

Clearly, I've offered you only a brief survey of the ideas of the second superstring revolution. The rest of this chapter, and much of chapter 6, will be devoted to developing some of these ideas more fully. The best place to start is D-branes.

D-branes and symmetries

D-branes are a particular type of brane. Their defining property is that they are locations in space where strings can end. It took a long time to realize that this simple idea can be developed into a remarkably rich understanding of how D-branes move and interact. D-branes have a definite mass which can be calculated starting just from the idea that strings can end on a D-brane. This mass becomes larger and larger when strings interact more and more weakly. A standard working assumption in worldsheet string theory is that string interactions are very weak. The D-branes are then so massive that it's hard to get them to move, and therefore hard to appreciate their role as dynamical objects in string theory. I suspect the prevalence of the assumption of weak string interactions prior to the second superstring revolution was another reason that it took a while for D-branes to be recognized as dynamical objects in their own right.

I introduced D0-branes in the previous chapter. They are point particles. D1-branes are like strings. They stretch out in one spatial dimension. They can close on themselves to form loops. And they can move in all sorts of ways, just like strings. That is, they can vibrate, and they have quantum fluctuations. A Dp-brane stretches out in p spatial dimensions. There are Dp-branes in 26-dimensional string theory, and also in ten-dimensional superstring theory. As I explained in chapter 4, 26-dimensional string theory has an awful problem: the string tachyon, which is a kind of instability. A similar instability afflicts D-branes in 26-dimensional string theory, but not in ten-dimensional superstring theory. Mostly, in the rest of this book, I'll be talking about superstring theory.

A lot can be understood about D-branes by understanding their symmetries. I've used the word symmetry pretty freely so far. Now let me explain what physicists mean by this word. A circle is symmetrical. So is a square. But a circle is more symmetrical than a square. Here's how I would jus-

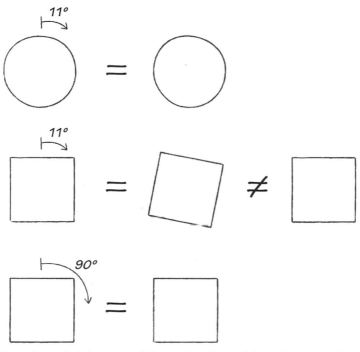

Rotating a circle by any angle leaves it unchanged. Rotating a square by 90° leaves it unchanged, but rotating it by other angles does change it.

tify that comparison. A square is the same if you turn it 90° degrees. A circle is the same no matter how you turn it. So there are more different ways of viewing the circle that make it look the same. That's what symmetry is all about. When something is the same when viewed from different angles, or in different ways, it has the quality of symmetry.

Physicists (and mathematicians) are deeply attached to a slightly more abstract description of symmetry. The key concept is called a group, or a symmetry group. When you turn a circle, let's say by 90° to the right, it corresponds to an

"element" of the group. This "element" is a rotation by 90°.
You don't have to be thinking of a circle to grasp the idea of a
90° rotation. Think of it this way. Everyone understands the
idea of a right turn. A right turn usually amounts to turning 90° to the right. We can talk about right turns without
discussing a particular intersection. We also understand that
a left turn is the opposite of a right turn. If you start going
north on 8th Avenue in Manhattan, make a right turn onto
26th Street, and then a left onto 6th Avenue, your direction
of travel is the same as it was when you started: north. I'll
admit that not everything is the same. You're now driving on
6th Avenue, whereas before you were on 8th. But suppose all
you're trying to keep track of is direction. Then, really, a right
turn and a left turn cancel out, just like 1 and −1 add up to 0.

There's another thing you know about right turns and left
turns—meaning turns by 90°. Three right turns amounts to
a left turn. After four right turns, you're moving in the same
direction you were going. This is very different from adding
and subtracting numbers. Think of a right turn as a 1, and a
left turn as a −1. Two right turns is $1 + 1 = 2$. Two right turns
and one left turn is $1 + 1 - 1 = 1$, so the same as one right
turn. So far so good. But four right turns is like not turning
at all, which would suggest $1 + 1 + 1 + 1 = 0$. Not good.
This illustrates the difference between the "arithmetic" of
right and left turns, and ordinary arithmetic. Mathematically,
all there is to know about a group is how its elements add
up. Well, not quite. You also have to know how to find the
"inverse" of a group element. The inverse of a right turn is a
left turn. Whatever a group element does, its inverse undoes.

There's a certain similarity between this discussion and
the one in chapter 4 about spacetime from strings. In that
section, we started by thinking of the string worldsheet as an

abstract surface. Then we told it how to move in spacetime. Here we're thinking of a group as an abstract collection of elements. Then we decide how those group elements act on a particular object, like a circle, or a square, or a traveling car.

I claim that the symmetry group of a square (more properly, the group of rotational symmetries of a square) is the same one that describes right turns and left turns. A right turn means rotating by 90°. When you're driving, turning right also means that you go around a corner: you turn at the same time as moving forward. But as I've said, we're trying to keep track only of the direction you're facing, not your forward progress. If that's all we're thinking about, then this turning by 90° is just a rotation, as if we stopped in the middle of the intersection, turned the car by some magical means, and then started going again. The point is that these 90° turns are exactly the ones that describe the rotational symmetries of a square. A circle is more symmetrical yet, because you can turn it by any angle and it's the same.

Is there anything more symmetrical than a circle? Sure: a sphere. If you take a circle and turn it out of the plane in which it lies, clearly it's not the same. But a sphere is the same no matter how you turn it. It has a bigger group of symmetries than a circle.

Now let's get back to D-branes. It's hard to keep track of ten or 26 dimensions, so let's just imagine that we've somehow done away with all but the usual four. A D0-brane has the symmetries of a sphere. Any point particle does—at the level of our current discussion. The reason is just that a point looks the same from any angle, just like a sphere. D1-branes can have many shapes, but the simplest to visualize is when it is perfectly straight, like a flagpole. Then it has the symmetries of a circle. If that doesn't make sense, think of a D1-brane rising straight up out of a sidewalk. OK, that was silly—think of a flagpole in the middle of a sidewalk. You can't really rotate the flagpole: it's too heavy. But you can look at it from different sides. It looks the same from all angles. The same is true of a circle drawn on the sidewalk. You can't turn it, but you can look at it from every angle, and it's the same.

Symmetry is an elaboration of the notion of sameness. So it may seem like it gets boring pretty quickly. Same old same old, eh? But there are a couple more elaborations that make it all seem a lot more exciting to me. First, think of a turntable. (For people younger than the author, it may help to be reminded that a turntable is the part of a record player that you put a record on.) If it's a really good one, with no wobbles, it is hard to tell by looking whether it is on or off. That's because it has the symmetries of a circle. But now imagine putting a record on it. You can tell that it's turning now because the center label usually has some words printed on it. But let's ignore that for now. The grooves on the record are in a spiral pattern. If you look closely, you can see that spiral moving. It looks like each individual groove is moving slowly, slowly inward. If you put a needle on the record, it follows the grooves inward. If you jiggered the turntable so that it turned backward, the needle would move slowly outward. The point is that continuous turning is not the same as staying still. We don't really need the record to tell us this: it just helped illustrate that rotational motion can be detected in obvious ways or in subtle ones. We could have just said that the turntable is continuously undergoing rotation and left it at that.

Particles like electrons and photons are forever rotating. The term physicists prefer is that they're spinning, like tops.

CHAPTER FIVE

Electrons can spin in any direction they like: that is, the axis of their rotation can point in any direction. Physicists commonly refer to the axis of rotation of a spinning electron as the direction of its spin. This axis of rotation can itself change over time, but it does so only under the influence of electromagnetic fields. Atomic nuclei spin in essentially the same manner as an electron. Magnetic resonance imaging (MRI) takes advantage of this. With an intense magnetic field, an MRI machine aligns the spins of the protons in hydrogen atoms in the patient's body. The machine then sends in a radio wave that tips the axes of some of the protons' spins. As the spins come back into alignment, they emit some additional radio waves. These emitted radio waves can be thought of as echoes of the ones that the MRI machine sent in. With a lot of sophistication and experience, physicists and doctors have learned how to "listen" to those echoes and figure out what they tell about the tissue that produced them.

Photons spin too, but not in just any direction. The axis of their rotation has to be aligned with the direction of their motion. This restriction cuts to the heart of modern particle physics, and it's a consequence of a new kind of symmetry, called a gauge symmetry. The word "gauge" refers to a system of measurement or a measurement device. For example, a tire pressure gauge is a device for measuring tire pressure, and the gauge of a shotgun is a way of stating the diameter of the barrel. In physics, when an object can be described in several different ways, and there's no *a priori* reason to prefer one over another, a gauge is a specific choice of which description to use. Gauge symmetry refers to the equivalence of different gauges. Gauges and gauge symmetries are pretty abstract notions, so let's consider a commonplace analogy before going further. I remarked earlier that it's hard to tell whether a turntable is turning or not, because it's symmetrical. A convenient way to remedy this is to mark the edge of the turntable with a dot of whiteout. It doesn't matter where on the edge you put the whiteout: For instance, you could put it on the side nearest you, or you could reach across and mark the other side. Wherever the mark is, its motion lets you tell at a glance that the turntable is turning. The choice of where to put the mark is like a choice of gauge. The arbitrariness of where you decide to put the mark is like gauge symmetry.

Gauge symmetry has two important consequences for the quantum mechanical description of photons. First, it ensures that the photon is massless, so that it always travels at the speed of light. Second, it restricts the axis of the spin to be always aligned with the direction of motion. It's hard for me to explain how these two restrictions arise from gauge symmetry without delving into the mathematics of quantum field theory. But what I can do is to explain the relation between them. Consider first an electron, which has both mass and spin. If the electron is at rest, it wouldn't make sense to say that its spin must be aligned with its motion, simply because it isn't moving. A photon, on the other hand, must always move at the speed of light. You can't move without moving in some direction. Thus it at least makes sense to restrict the axis of a photon's spin to be aligned with the direction of its motion. In short, the first restriction (masslessness) is necessary in order for the second (spin alignment) to make sense.

The consequences of gauge symmetry make it seem like a very different idea from the symmetries we discussed previously. It's more like a set of rules. A photon can't stand still, because of gauge symmetry. It can't spin in certain directions, because of gauge symmetry. There's one more important thing to know: electrons have charge because of gauge sym-

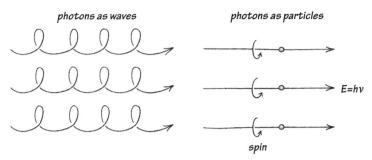

Photons as waves and as particles. In the particle description, the axis of
the spin is aligned with the motion. In the wave description, the electric
field has a corkscrew shape. If all the photons spin the same way, as I
drew, the light is described as "circularly polarized."

metry. An analogy between gauge symmetry and the rotational symmetry of a turntable helps illustrate this last point.
Gauge symmetry for an electron is like rotational symmetry:
one even speaks of "gauge rotations." But a gauge rotation
isn't a rotation in space. It's more abstract, and it relates to the
way one describes electrons quantum mechanically. Just as a
turntable rotates at a constant rate (when it's turned on), an
electron "rotates" in a more quantum mechanical sense, related to gauge symmetry. This rotation is its electric charge.
Electric charge is negative for electrons and positive for protons. That means they "rotate" in opposite directions in the
abstract sense that relates to gauge symmetry.

It turns out that extra dimensions help make the whole
discussion of charge more concrete. If there's one extra dimension, and its shape is a circle, then you can imagine a
situation where a particle is going around the circle. It can go
around forward or backward. If the circle is really small, you
would not notice it like the usual four dimensions. Nevertheless, elementary particles could go around the circle, forward

or backward. If they went forward, they would have positive charge. If they went backward, they would have negative charge. The whole setup relies on a circular extra dimension, so it shouldn't surprise you to learn that the symmetries of a circle have a lot to do with gauge symmetry. In fact, the gauge symmetry of electric charge is the same as the symmetry of a circle. That may seem like an abstract statement. But it has an immediate consequence. Motion in a circle is either forward or backward. There isn't any other direction. In the same way, electric charge is either positive or negative. There isn't any other kind of charge.

The idea of explaining electric charge in terms of a circular extra dimension predates string theory. It is a little less than a hundred years old. However, it's never been made to work quantitatively. Part of the grand ambition of string theory is to make this idea come to fruition. We certainly have a bunch of extra dimensions to play with, so there should be some hope. Whether we're on the right track or not with extra dimensions, the idea of gauge symmetry is here to stay. Electric charge and its interactions are fundamentally connected to the symmetries of a circle and motions around a circle.

It seems we have wandered far from D-branes. But not really. D-branes provide examples of everything we've discussed. We've already seen how D-branes have rotational symmetries: recall the comparison between a D1-brane and a flagpole, whose rotational symmetry is the same as the symmetry of a circle. Rotational symmetries help explain the properties of D-branes. But gauge symmetry plays a big role too. Here's the first hint of the connection between gauge symmetry and D-branes. If you start with a D1-brane that is stretched out straight, and you tap it in a particular place, two little ripples will move out from where you tapped it.

CHAPTER FIVE

These ripples will move at the speed of light. They are like massless particles. Nothing will make them stand still. I've explained that massless particles like photons are associated with a gauge symmetry, and that their property of being massless is enforced by the gauge symmetry. That's essentially what's going on with the ripples on a D1-brane. I'm oversimplifying, because these ripples aren't quite like photons. They have no spin. But, if we were to discuss ripples on a D3-brane, then some of them do have spin, and they have exactly the same mathematical description as photons. Pretty much as soon as D3-branes were invented, people started trying to build a model of the world in which the dimensions we experience are the ones on a D3-brane. There are still extra dimensions, but we can't get at them because we're stuck on the brane. What seems to give this idea a chance is that D3-branes come equipped with photons. All we need is the other fifteen or so fundamental particles, and we'd be set. Sadly, a D3-brane by itself doesn't provide them. It's an active area of research to find out what other ingredients you'd need in order to build the world on a D3-brane.

D-branes in superstring theory also have charge, similar to electric charge. The analogy is quite precise in the case of D0-branes. They have a charge that we could say is +1. There is another object, an anti-D0-brane, that carries charge –1. Now, remember our discussion of the almost hundred-year-old idea that charge is associated with a circular extra dimension? It works perfectly for D0-branes. One of the breakthroughs of the second superstring revolution was that superstring theory was concealing an extra dimension, beyond the ten that we were used to. A D0-brane, which you'll remember looks like a point, can be described as a particle moving around that eleventh dimension, which is

84

rolled up in a circle. If a particle moves the other way around the eleventh dimension, it's an anti-D0-brane. This realization is what made people suddenly take eleven-dimensional supergravity seriously. In some sense, string theorists were studying it all along without realizing it! And it turns out that the eleventh dimension doesn't have to be curled up in a small circle. As you make the circle bigger and bigger, interactions between superstrings get stronger and stronger. They split and join so rapidly that it seems hopeless to try to keep track of them. But as the dynamics of the string picture becomes more complicated, a new dimension literally opens up. Eleven-dimensional supergravity becomes the simplest description of the strongly interacting superstrings. We don't know exactly how to merge quantum mechanics with eleven-dimensional supergravity. But we feel convinced that there must be some way to do it, because string theory is a fully quantum mechanical theory, and it clearly includes eleven-dimensional supergravity when the superstring interactions become strong. It is this circle of ideas that soon got the name of M-theory.

A great hope of string theorists is that all our notions of charge and gauge symmetry might simply stem from the secret higher-dimensional nature of the world. In chapter 7 I'll discuss more fully how this might work. In chapters 6 and 8, I'll explain how extra dimensions might be used to describe the strong interactions, like the interactions among quarks and gluons inside the proton. To give you a brief preview: In some circumstances, or in some approximation, these interactions might be effectively described in terms of a fifth dimension. This fifth dimension "opens up," like the eleventh dimension of M-theory, when interactions become too strong to keep track of in the usual four dimensions.

CHAPTER FIVE

D-brane annihilation

As I explained in the previous section, D0-branes carry a charge, and there is another object, called an anti-D0-brane, that carries the opposite charge. What would happen if a D0-brane collided with an anti-D0-brane? The answer is that they would annihilate each other, disappearing in an explosive burst of radiation. This section is devoted to describing in more detail how D0-branes and anti-D0-branes interact.

To get started, let's go back to the discussion in chapter 4 of strings stretched between D0-branes. The purpose of that discussion was to tell you about the three contributions to the mass of a string. There was the rest mass, which came from stretching the string between the branes. There were the vibrations, which were like the motion of a piano string when struck. And there was the contribution of quantum fluctuations, which was negative, and very hard to get rid of. And very problematic, because it led to tachyons—things with *imaginary* mass. I mentioned that one way to get rid of tachyons was to move the D0-branes far enough apart so that the stretching energy was bigger than the negative contribution from quantum fluctuations. Well, let's turn this around. What if we start with D0-branes far apart, and then bring them closer and closer together? The answer depends on details. To get the story straight, we have to distinguish carefully between D0-branes and anti-D0-branes. The only difference between them is their charge. Consider first the case of two D0-branes approaching one another. They have the same charge. That means that they repel one another the way electrons do. But they also have mass, so they exert a gravitational attraction on one another. The overall attraction precisely cancels out the repulsion. The upshot is that they scarcely notice one another. And it turns out that superstrings stretched between two D0-branes never turn into tachyons. This is a small example of the miraculous solution of the tachyon problem in superstring theory.

Everything changes when you consider a D0-brane close to an anti–D0-brane. The D0-brane and the anti–D0-brane have opposite charges. So they attract one another, like an electron and a proton. The gravitational attraction is unchanged, just because D0-branes and anti–D0-branes have the same mass, and gravity responds to mass. The upshot is that there's a strong attraction between a D0-brane and an anti–D0-brane. Strings stretched between them know about this attraction. The way they know about it is that they become tachyons when the D0-brane and anti–D0-brane get too close. I noted in the previous chapter that the modern understanding of a tachyon is that something is unstable. The example I gave was a pencil standing on its point. Eventually, it falls over. A D0-brane sitting right on top of an anti–D0-brane is similarly unstable. What happens, as I remarked at the beginning of this section, is that they mutually annihilate. The annihilation process is analogous to the pencil falling over. You can get an alternative view of it by thinking of the eleventh dimension in the shape of a circle. A D0-brane is a particle going around the circle. An anti–D0-brane is a particle going the other way. If the D0 and anti–D0 are right on top of one another, the particles will collide. When they do, the D-branes disappear in a flash of radiation. The details of this process should teach us something about M-theory, but unfortunately it's not very well understood. The trouble is that the annihilation process is very rapid, and it's hard to track the way in which a large amount of energy gets released in a short time. What we can be sure of, based on $E = mc^2$,

CHAPTER FIVE

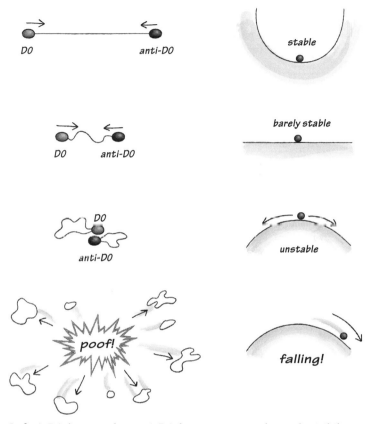

Left: A D0-brane and an anti-D0-brane come together and annihilate into strings. A string stretched between them becomes tachyonic when the branes get too close. Tachyonic means unstable. The tachyon is a quantum of instability. Right: When the D0 is far from the anti-D0, the would-be tachyon is actually stable. When the D0 and anti-D0 come too close, the tachyon rolls away. This rolling is equivalent to the annihilation of the D0 and anti-D0.

BRANES

is that the energy released is twice the rest energy of the D0-brane, plus any kinetic energy that the D0 and anti-D0 may have had before the annihilation.

Branes and black holes

I introduced D-branes as locations in spacetime where strings are allowed to end. It turns out that there is another way to think about them: they are zero-temperature black holes. This way of thinking is best when there are many D-branes on top of one another. Let's start with D0-branes. As I just explained in the previous section, in superstring theory, two D0-branes don't exert any net force on one another. Their gravitational attraction is canceled by their electro-static repulsion, and they don't annihilate each other like a D0-brane and an anti-D0-brane do. So we can consider two D0-branes on top of one another, or indeed any number, without worrying about violent processes like annihilation. However, the more D0-branes you have, the more the nearby spacetime is distorted around them. The distortion takes the form of a black hole horizon. To make this seem more plausible, think of a million D0-branes sitting on top of one another, and one lonely D0-brane moving nearby. The lonely D0-brane is neither attracted nor repelled. Actually, there's a caveat to that statement. The lonely D0-brane doesn't feel any net force at all if it isn't moving. If it is moving, then it does feel just a little tug toward the other branes. Similar tugging helps prevent the million D0-branes from dispersing. Everything is quite different for an anti-D0-brane. It feels both gravitational and electrostatic attraction, just like I described before. When it gets really close to the big clump of a million D0-branes, it's like one of those fish in the lake who ventured too close to the drain. It gets sucked in. No physical process can save it when it gets closer than a certain distance. That's basically the notion of a black hole horizon.

What about the claim that the horizon has zero temperature? This is harder to explain. It has to do with the behavior of the lonely D0-brane, which feels no net force from the clump. It turns out that this no-force condition is deeply related to zero temperature. Both properties are enforced by supersymmetry. I defer a careful discussion of supersymmetry to chapter 7, but let's add incrementally here to our familiarity with supersymmetry with the following two statements. First: Supersymmetry relates gravitons and photons. Gravitons control gravitational attraction. Photons control electrostatic attraction or repulsion. The particular relation that supersymmetry implies between gravitons and photons just says that gravitational and electrostatic forces are equal. Second: Supersymmetry guarantees that D0-branes are stable. What this means is that there's no lighter object in string theory that a D0-brane can turn into—unless it encounters an anti-D0-brane. So a D0-brane, though heavy, is quite *unlike* a uranium-235 nucleus, which can decay into lighter nuclei such as krypton and barium, as I discussed in chapter 1.

Clumps of D0-branes are also stable. They can't decay into anything else. The only thing they can do when they are together is vibrate a little. These vibrations are like the thermal vibrations of atoms in a lump of coal. If you remember, thermal vibrations can be translated into energy according to the formula $E = k_B T$. Here E is the extra energy due to the thermal vibrations. For example, when you apply this formula to a carbon atom in a lump of anthracite coal, E is the extra energy of the atom due to its thermal vibrations, not its rest energy. The total energy of the lump of coal should

BRANES

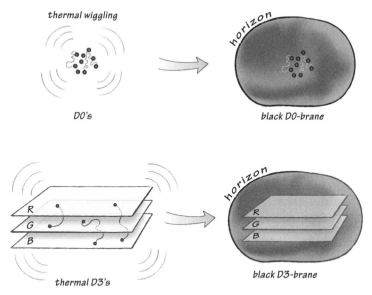

Top left: A clump of D0-branes with thermal energy. Top right: A horizon forms around the D0-branes to describe their thermal properties. Bottom left: Three D3-branes on top of one another. Strings running between the branes act like gluons and can provide thermal energy. Bottom right: A horizon forms around the D3-branes to describe their thermal properties.

include the rest energy of all the atoms and also the energy of their thermal vibrations. The atoms also have some quantum fluctuations in their positions, and in principle that also gets included in the total energy of the coal. It's all pretty similar to our earlier discussion of the three contributions to the mass of a string. The total mass of the lump of coal can be figured from its total energy using $E = mc^2$.

Now, all this discussion of coal can be carried over to the clump of D0-branes. They've got rest mass, and they've got some quantum fluctuations. In the case of D0-branes, the quantum fluctuations make exactly zero contribution to the total mass. (It's always a headache to track those quantum fluctuations!) The D0-branes might have some thermal fluctuations too. If they do, then the clump of D0-branes has a temperature, and it has extra mass. But it doesn't have extra charge. Now, if that lonely D0-brane happens to be close to a clump of D0-branes at some non-zero temperature, then the extra mass is going to add a little extra oomph to the gravitational pull on the lonely D0. So it will be drawn in. If you cool off the clump of D0-branes to absolute zero, it loses that extra bit of mass. So it no longer exerts any force on the lonely D0-brane. This is how zero temperature is associated with a no-force condition.

If all this talk of D0-branes is losing you, let's take a break and talk about coal some more. Its thermal vibrations get included in its total energy, just as for the clump of D0-branes. This total energy is still the energy of the coal at rest. "At rest" just means that the coal is sitting there, as opposed to flying through the air. Total rest energy gets translated into total mass via $E = mc^2$. So the lump of coal is actually heavier when it's hot than when it's cold, just as a clump of D0-branes is heavier when it's hotter. In the case of a lump of coal, you can plug in some everyday numbers and figure out just how much mass the coal gains from being hot. Here's how I would do it. A really hot coal is about 2000 Kelvin. If you remember, the surface of the sun is only about three times hotter. $E = k_B T$ is an estimate of the thermal energy in each atom of coal—but only an estimate. Using this estimate without trying to improve on it, I calculated that the thermal energy in a hot coal is about 10^{-11} times its rest mass. That's one part in a hundred billion. This is much more than the fraction of rest mass that an Olympic sprinter can turn into

BRANES

kinetic energy in the hundred-meter dash. But it is much, much less than the fraction of rest mass converted to energy in nuclear fission. That's basically why nuclear power is so promising: a ton of fuel-grade uranium, used in a modern nuclear reactor, yields about the same amount of electrical energy as a hundred thousand tons of coal.

I doubt you'll be happy to hear it, but the discussion of D0-branes was oversimplified in two ways. First, there's another interaction among D0-branes, mediated by a massless particle that is neither a photon nor a graviton. This particle is called a dilaton, and it has no spin. Everything I said about gravitational attractions should really have been extended to include the dilaton. But even with that little change, the final conclusions are the same. Second, if D0-branes are behind a horizon, it's hard to say whether they're vibrating like atoms do. All you can say for sure is that the clump of D0-branes has some extra energy, which is the same as extra mass. A big problem in string theory is to give a more precise description of black holes made out of vibrating D-branes. The best-understood case involves D1-branes and D5-branes. Another important case is D3-branes. The D0-brane case is harder to work out quantitatively, but there has nevertheless been significant progress.

When we switch from discussing the black hole view of D0-branes to D1-branes or D3-branes, the main thing that changes is the shape of the horizon. D3-branes surrounded by a black hole horizon are hard to visualize because the D3-branes extend over three spatial dimensions. You have to visualize at least one more dimension to get a proper idea of what the horizon looks like. In later chapters, I'm going to work on explaining this case some more, because it's really the most interesting. For now, let's consider D1-branes in the four spacetime dimensions of everyday experience—assuming, as

CHAPTER FIVE

we did before, that we somehow did away with the other six. When a single D1-brane is stretched out straight, it looks like a flagpole, and its fluctuations are the ripples I described earlier (see p. 82). When many D1-branes come together, there are more types of ripples. The best way to think of these ripples is in terms of strings. A string can have one of its ends on one D1-brane, and its other end on another. It can slide along the D1-branes in the direction they've been stretched out. A string with ends is generally called an open string. That's in contrast to a closed string, which by definition is a closed loop. Adding thermal energy to D1-branes basically means adding open strings. Surprisingly, the open strings describe all possible small vibrations of the D1-branes. In other words, the strings are essentially the ripples on the D1-branes.

If there are a large number of D1-branes, then the whole assembly of D1-branes and open strings on them distorts the nearby spacetime, and a black hole horizon forms. The horizon has the symmetries of a circle, just as a single stretched D1-brane does. You can think of the horizon as a cylinder that surrounds the assembly of D1-branes and open strings. That's a different shape from the horizon around a clump of D0-branes, which is spherical. Some string theorists prefer to use the term "black brane" to describe a group of D1-branes surrounded by a horizon. They reserve the term "black hole" for a spherical horizon, like the one surrounding D0-branes. I favor a slightly looser usage: black branes, black holes, whatever rolls off the tongue more easily. For example, I would describe the cylindrical horizon surrounding a group of D1-branes as a black hole horizon, and I would refer to the whole geometry as a black D1-brane.

Historically, it's interesting to note that the black hole (or black brane) geometries describing clumps of D-branes were known before D-branes themselves were properly understood. To understand black branes, all you need is to be able to solve the equations of supergravity. If you recall, supergravity is a low-energy limit of superstring theory, where you forget about all the overtones of string vibration and focus in on vibrational modes that are massless. Supergravity is still pretty complicated. But it's a lot simpler than superstring theory. The construction of black branes is one of several ways in which supergravity helped guide the development of string theory during the second superstring revolution.

Branes in M-theory and the edges of the world

So far, I have focused my discussion of branes on D-branes. I did this because D-branes are the most important, best understood, and most diverse collection of branes. But it would be a distortion to leave out the other branes. Partly this is because they are weirder than D-branes. There's probably more left to be discovered about them. Weirdest of all are the branes of M-theory.

M-theory, I remind you, is the quantum mechanical theory that includes eleven-dimensional supergravity as its low-energy limit. Although M-theory is more than ten years old at the time of writing this book, the statement I've just made is still the most important thing we know about it. I would not hesitate to say that this is a disappointment. Still, there's a lot to eleven-dimensional supergravity. In particular, it includes two black branes: the M2-brane and the M5-brane. They are similar to the black branes in string theory that describe groups of D-branes surrounded by a horizon. They are especially similar to the black D3-brane.

CHAPTER FIVE

M2-branes are extended in two spatial directions, and M5-branes in five. Like D-branes, they can be stretched out straight in the eleven dimensions of M-theory, or they can wrap around and close on themselves. Unfortunately, we don't understand very well how M-branes fluctuate. We can track the motions of a single M2-brane that's stretched out and nearly flat. Its motions are like the ripples on D1-branes that I described in the previous section. We can similarly track the motions of a single M5-brane. But when multiple M-branes sit on top of one another, the story becomes more complicated, and it has defied understanding for many years. Literally as I write this chapter, this wall of ignorance seems to be cracking. A handful of papers have appeared that purport to describe the dynamics of two or more M2-branes on top of one another. But we are still far, far away from the level of detailed understanding we have in the case of strings. We can keep track of how a string moves, both classically and quantum mechanically, whether the string is nearly straight or flopping all over the place. There are still some conceptual roadblocks to a similar understanding of M2-branes. And M5-branes are, if anything, even more mysterious.

There's one further type of brane in M-theory that is really surprising. This brane is the edge of spacetime. It's where space itself ends. Usually, in string theory, space can't end, any more than a string can end without a D-brane around. The space-ending brane is one of the wilder ideas in M-theory, but it's actually very well accepted. It turns out that there are photons at the edge of spacetime, much like the photons on D-branes. But the photons at the edge of spacetime participate in a particularly interesting theory called supersymmetric E_8 gauge theory. A lot of work in the middle 1980s, after the first superstring revolution, revolved around folding this theory up in such a way as to recover the theories of electromagnetic and nuclear forces. It turns out that all this work has an M-theory interpretation in terms of spacetimes that end on a space-ending brane.

The space-ending brane is one of the ways in which M-theory has moved decisively beyond eleven-dimensional supergravity. This advance required some use of quantum mechanics. Another such advance was the calculation of the mass of M2-branes and M5-branes. Actually, the mass of an M2-brane is infinite when it is stretched out straight and flat across an infinite area. The same goes for M5-branes. What's understood, again from quantum mechanics, is that the mass per unit area of an M2-brane is a fixed number. This is actually more information than we have about string theory itself: the mass per unit length of a string is arbitrary as far as we know.

Besides D-branes and M-branes of various stripes, there's one more brane in superstring theory. Actually, it was the first one to be understood. It's a 5-brane, like the M5-brane, but it lives in ten dimensions, not eleven. It's sometimes called the solitonic 5-brane, and for lack of a more descriptive name, I'll stick with that one. Solitons are a widespread notion in physics, and in general they are heavy, stable objects. The classical example is a wave that can travel along a channel, like a canal, without dissipating or breaking. "Soliton" evokes the word "solitary." It's supposed to communicate the idea that a soliton has its own identity. Nowadays we understand that D-branes have their own identity too. All branes can be loosely described as solitons of string theory. But here I'll use "solitonic" to describe only the 5-brane I was just starting to talk about.

The solitonic 5-brane is worth mentioning for two reasons. First, when we get to discuss string dualities, it's useful to know that the solitonic 5-brane exists, because duality symmetries relate it to other branes. Second, our understanding of the solitonic 5-brane is an example of the idea that spacetime has no meaning by itself, but exists only to describe how strings move. I tried to illustrate this idea in chapter 4 using an analogy between strings in spacetime and cars on a racetrack. The first salient feature of a racetrack that I suggested could be deduced from a record of how the racecars move is that it's a closed loop. Well, the central idea of the solitonic 5-brane is similar. You start out by assuming that superstrings move on the surface of a sphere. Actually, for technical reasons, the sphere you use has one dimension more than the one that approximates the shape of the Earth's surface. This higher-dimensional sphere is called a 3-sphere. What I want to convey is that it's like the racetrack in my analogy: closed, finite, and of definite size. Now, if you recall, superstrings are quite picky about what kind of geometries they will tolerate. They insist on ten dimensions, and they insist that the equations of general relativity should be obeyed. Having started with a 3-sphere, you need to add time plus six spatial dimensions. The overall shape you wind up with is quite distinctive. Here's how it looks. Spacetime far from the solitonic 5-brane is flat and ten-dimensional. As you move inward, you find a deep hole in spacetime with a definite size: the size of the 3-sphere you started out with. This "deep hole" is related to a black hole, just like every other brane in string theory. But it turns out you can go as deep into a solitonic 5-brane as you like without crossing a horizon. What that means is that no matter how deep you go into a solitonic 5-brane, you can always turn around and come back. Physics deep down in the hole eventually gets pretty strange: strings start interacting strongly, and in

some cases an extra dimension opens up, bringing us back
to eleven dimensions.

I hope this chapter leaves you with two overall impressions.
First, strings are not the whole story—far from it. Second,
the whole story is complicated and detailed. At least, it seems
complicated and detailed. Often, when things get so complicated and detailed, a deeper level of understanding eventually simplifies the story. A good example is chemistry, where
there are about a hundred different chemical elements. The
unifying understanding came from the realization that all of
them are made up of protons, neutrons, and electrons. There
is a similar profusion of elementary particles in the conventional understanding of high-energy particle physics. There
are photons, gravitons, electrons, quarks (six kinds!), gluons,
neutrinos, and a few others. String theory aims to be a unifying picture, where each of these particles is a different vibrational mode of a string. At some level it's disappointing to
learn that superstring theory has its own proliferation of different objects. On the positive side, this proliferation forms
an extraordinarily tightly woven web, where every type of
brane can be related to every other one and to strings. These
relations are the topic of the next chapter.

It's hard to keep from wondering whether there's something deeper and simpler than branes—maybe some kind of
"sub-brane" from which all branes are composed. I don't see
any hint of "sub-branes" in the mathematics of string theory.
But there are certainly hints in plenty that our understanding of that mathematics is incomplete. The third superstring
revolution, if it ever comes, has a lot of problems to solve.

CHAPTER FIVE

Chapter S I X

STRING DUALITIES

A DUALITY IS A STATEMENT THAT TWO APPARENTLY DIFFER-
ent things are equivalent. I already discussed one example
in the introduction: a checkerboard. You can think of it as
black squares on a red background, or as red squares on a
black background. Those are "dual" descriptions of the same
thing. Here's another example: dancing the waltz. Probably
you've seen this in old movies, or maybe you've even done
it. The man and the woman face one another, close together.
There's a particular way you hold your arms, but never mind
about that. What matters most is the footwork. When the
man steps forward on his left foot, the woman steps back on
her right foot. When the man steps forward on right, the
woman steps back on left. As the man turns, the woman
turns to stay facing him. If you leave out fancy moves like
spins, you could work out exactly what the woman should be
doing based on what the man does—and vice versa. There's
the old joke that Ginger Rogers did everything that Fred
Astaire did, but backwards and in heels. That's kind of like a

string duality. Every object in one description can be equally well captured in another.

When you watch Fred and Ginger dance in an old movie, part of the charm of the dance is how they mirror one another. Similarly, in string theory, when you understand a duality, you get a more insightful and informative picture than if you only understood one side of the duality. Limiting yourself to only one side of the duality would be like watching only Fred, or only Ginger. Captivating, maybe, but incomplete.

Here's a real example of a string duality. We've talked about strings, and we've talked about D1-branes. Both of them extend in one dimension of space. As in the previous chapter, I mostly want to focus on ten-dimensional superstring theory, as opposed to the 26-dimensional string theory that has tachyon instabilities. A famous string duality, called S-duality, interchanges superstrings with D1-branes. That's interesting, but it's only one aspect of the duality—as if all I told you about the waltz is that the woman steps back on her right foot when the man steps forward on his left. In order to give a fuller account, I have to tell you what S-duality does to every brane in superstring theory. Before I do that, I have to introduce one additional complication. There are different kinds of superstring theory. They can be distinguished by which kinds of branes are allowed. The type of superstring theory I want to talk about is called Type IIB. This name is not very descriptive. It was chosen before a lot of the unique dynamics of this particular string theory were understood. But I'll stick with it. Type IIB string theory has D1-branes, D3-branes, D5-branes, solitonic 5-branes, and a few other branes that are more complicated to explain. It doesn't have D0-branes, or D2-branes, or any other even-numbered brane. It's a string theory, not M-theory, so it doesn't have M2-branes or M5-branes.

CHAPTER SIX

Back to S-duality. I introduced it by saying that strings are exchanged with D1-branes. It turns out that D5-branes are exchanged with solitonic 5-branes, and D3-branes are unaffected by the duality. What this means is that if you start with a string on one side of S-duality, you end up with a D1-brane on the other side; but if you start with a D3-brane on one side, you end up with a D3-brane on the other. There's more to the story, but already we can put together some of the statements I've made to learn something new. A string can end on a D5-brane. (This is because a D5-brane, like any D-brane, is defined as a location where strings can end.) How does S-duality affect this statement? S-duality instructs us to replace "D5-brane" with "solitonic 5-brane" and "string" by "D1-brane." So the new statement is that a D1-brane can end on a solitonic 5-brane. This new statement can be independently checked, and it's true. String dualities got built up in roughly this way: certain rules of translation were proposed, then new consequences were deduced and checked.

In general, a string duality is a duality relation between two apparently different string theories, or between two apparently different constructions in string theory. A whole web of string dualities is now known. It is so well connected that you can start with any brane you like, go through a few dualities and "deformations," and wind up with any other brane. I'll explain as I continue just what I mean by a deformation. Before we start, it's worth returning to an earlier point about unifying pictures that we discussed near the end of chapter 5. There are so many different branes in string theory! One might expect eventually to find a unifying picture where all branes are different manifestations of the same underlying structure. Dualities aren't like that. They trade one type of brane for another. Sometimes they trade branes for strings. At our current level of understanding, it appears that all types of strings and branes are at some level coequal. This is qualitatively more than chemists understood about the different elements of the periodic table before atomic theory. But it is qualitatively less than physicists understood about chemical elements after atomic theory was well established.

The story of string dualities got going just as I was a beginning graduate student. I remember viewing them with some skepticism. Was this really what I wanted to study? It certainly was a pretty subject, but it seemed pretty distant from the goal of making string theory into a theory of everything. My take on the subject now is that it was an inevitable progression in our understanding. Some of the best prospects for connecting string theory to experiment rest squarely on dualities.

Our understanding of string dualities varies. S-duality is actually one of the more mysterious kinds of dualities. The rule for mapping strings to D1-branes is well understood and well checked in the case where the strings or the D1-branes are stretched out, straight, and (nearly) motionless. But the rules of S-duality are not so well understood for strings or D1-branes that are flopping around and colliding with one another in arbitrary ways. The difficulty relates to the strength of string interactions. I've described the splitting of a string into two strings as similar to a pipe branching into two smaller pipes. The surface of the pipe is like the worldsheet of the string, which is the surface in spacetime that the string sweeps out over time. Strings joining would be like two pipes coming together into one larger pipe. The strength of string interactions is a way of quantifying how frequent those splitting and joining events are. When string interactions are weak, a string travels a long way before splitting or interacting with another string. When string interactions are strong, there are so many splittings and joinings that you can scarcely keep track of a single string: no sooner would you identify it than it would split, or join another string. When strings interact strongly, D1-branes interact weakly, and vice versa. So S–duality interchanges weakly interacting and strongly interacting behavior.

In case all of this got out of hand for you, let's return to the dancing analogy. Weakly interacting behavior in string theory is clean, simple, and elegant. It's like Fred Astaire's dancing. Strongly interacting behavior is chaotic and messy. Strings fly all over the place, but they're hardly strings anymore because they're splitting and joining so fast. The only analogy I can think of is a slimy alien. So S–duality is like Fred Astaire dancing with a slimy alien—sorry, Fred. But this alien is actually just as good a dancer as Fred, in its own way. We just can't easily appreciate what it's doing. If we were aliens ourselves, the opposite would be true. We'd perceive the alien's dance as clean, simple, and elegant, and because of our altered perspective, it would be Fred that looked like a slimy mess. The point that I'm aiming at in this analogy is that string dualities often relate something that we understand well (like weakly interacting string theory) with something that we don't (like strongly interacting behavior).

You may recall that when I discussed strongly interacting string theory in the previous chapter, the upshot was that a new dimension opened up. I claimed that string theory starts behaving like it is actually eleven-dimensional, not ten-dimensional. That's pretty different from what I explained in the last few paragraphs. In fact, I had in mind a different string theory. The one that grows an extra dimension when the string interactions become strong is called Type IIA string theory. It has D0-branes, D2-branes, D4-branes, D6-branes, solitonic 5-branes, and some other objects that are a bit harder to explain. When the string coupling is strong, Type IIA string theory is best described in terms of eleven dimensions. But Type IIB string theory at strong coupling is best described by swapping D1-branes for strings, without doing anything funny with extra dimensions.

I've emphasized already that there's a lot we don't understand about string dualities. So it's worth closing this section by summarizing the two things that we do understand reliably for every single string duality. The first is the low-energy theory. In every string theory we know, gravity is always part of the story. The description of gravity in general relativity is very special and very durable. It has a limited set of generalizations, which are the supergravity theories that I mentioned in the previous chapter. Supergravity theories capture the low-energy dynamics of superstrings because they include only the lowest energy vibrational modes of the superstrings. Our understanding of gravity and supergravity is so complete that, collectively, they become one of the main touchstones for our understanding of string dualities. The second touchstone is long, straight strings and long, straight branes. These are the objects that can be described as zero-temperature black holes in supergravity. They also have special no-force properties, such as those I described in my discussion of D0-branes. A minimum specification of a string duality amounts to describing what happens to the low-energy theory, plus what happens to these long, straight branes.

A dimension here, a dimension there, who's counting?

In this section, I want to discuss the best-understood string duality. It's called T-duality. These names—S-duality and T-duality—are as arbitrary as Type IIA and IIB. String theorists face a peculiar difficulty in naming things: we explore at the edge of knowledge and have to have names for things. So we make them up as we go. Often, the names we pick are arbitrary, or they refer to some very early work on a topic. But the names tend to stick even if the relevance of the early work fades. So we wind up with a hodge-podge of funny names. I think other fields of science have similar difficulties, but maybe not to the same degree.

Anyway, T-duality is the string duality that relates Type IIA and IIB string theory. It is so well understood because the whole story makes sense when strings interact only weakly. That means that strings travel a long way, or last a long time, before splitting or joining.

There is obviously a big problem in relating Type IIA and IIB string theory. IIA string theory has even-numbered D-branes: D0, D2, D4, D6. IIB has odd-numbered D-branes: D1, D3, D5. How can you possibly map a D0-brane onto a D1-brane? Especially if the D1-brane is long and straight, it seems impossible. Well, here's the trick. You roll up one of the ten dimensions of type IIA string theory on a circle. If that circle is much smaller than the length scales you can observe, then it looks like string theory has only nine dimensions. You could keep rolling up more dimensions until you get down to four. But let's not do that. We're trying to explain the relations among string theories here, not (at least, not yet) their possible relation to the world. So let's stick with just one rolled up dimension. In our new nine-dimensional world, the claim is that you can't tell the difference between Type IIA and Type IIB string theory. Take a type IIA D0-brane, for example. If you wrapped a D1-brane all the way around the circle, it would look like a D0-brane to an observer whose observational powers aren't precise enough to

STRING DUALITIES

see the size of the rolled up dimension. All I mean is that to such an observer, the wrapped D1-brane wouldn't seem like it had any spatial extent at all. It would seem like a point particle: a 0-brane. But wait! Isn't it possible for the D1-brane not to be wrapped, but instead to extend in one of the nine dimensions that our hypothetical hyperopic observer can see clearly? Well, yes, it is possible. On the other hand, it's also possible to wrap a D2-brane around that circular dimension. Then its shape is like a long hose. The cross-section of a hose is circular: that's the circular dimension that we rolled up. Just as a hose can snake across your lawn in a more or less arbitrary way, so a wrapped D2-brane can wander across nine dimensions. To the nine-dimensional observer we've been discussing, it looks just like a D1-brane. That's because this observer can't see closely enough to tell that the D2-brane is wrapped around on the extra dimension. The story continues about as you'd imagine: wrapped D3-branes act like D2-branes, wrapped D4-branes act like D3-branes, and so on.

The discussion so far might leave you with the impression that T-duality is only an approximate truth. Type IIA and IIB string theory look the same to a nine-dimensional observer only if she's not allowed to look so closely that she can discern the tenth dimension rolled up as a circle. Actually, T-duality is an exact duality. When you look at it using just the right mathematical language, it's almost as simple as the checkerboard duality between red and black squares. Although that mathematical language isn't really available to us, I can tell you the main point: a Type IIA string wrapped around the circular dimension is the same thing as a Type IIB string that isn't wrapped, but that instead is moving around the circle. Conversely, a Type IIA string that is moving around the circle is the same thing as a Type IIB string that is wrapped around it.

CHAPTER SIX

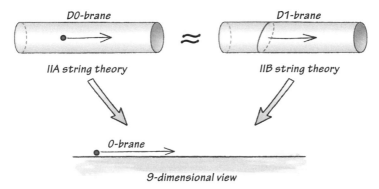

T-duality of Type IIA and Type IIB string theory. Both are related to a nine-dimensional theory. A 0-brane in nine dimensions can originate from a D0-brane in Type IIA theory, or equivalently from a D1-brane wrapping the circle in Type IIB theory.

The tricky bit is that the circle that the Type IIA string can wrap or move around has a different size than the one that the Type IIB string can move around or wrap. To understand this, we have to remember a little bit about quantum mechanics. When an electron moves inside an atom, it has definite, quantized energies, but its position and momentum are uncertain. A string that moves quantum mechanically around a circle is similar: it too has definite, quantized energies, but uncertain position. It turns out that the momentum of the string is quantized, like the energy. This is interesting, because it means that the uncertainty principle doesn't apply in its usual form to motions on a circular dimension. Instead, the mathematics that leads to the uncertainty principle tells us that when the circle is very small, the momentum of a moving string must be very big. As a consequence, its energy is very big too. Conversely, if the circle is very big, then the energy of a moving string can be very small. Let's compare

STRING DUALITIES

this situation with the energy of a string wrapping around a circle. The mass of a wrapping string is proportional to its length: that is, if you double the length, you double the mass. This is one way in which a string in string theory behaves just like ordinary string: it has a fixed mass per unit length. It follows that a string wrapping once around a big circle must be very heavy, while a string wrapping once around a small circle is light. Now we can come to the punch line. If you're going to replace a Type IIA string that is moving on a circle with a Type IIB string that wraps around a circle, you'd better do it in such a way that the energies match. If the circle that the Type IIA string theory moves on is small, then the energy is big, and that means that the circle the Type IIB string wraps had better be big. Likewise, if the IIA circle is big, the IIB circle must be small. If you squeeze the IIA circle smaller and smaller, the IIB circle gets so big that you can scarcely tell that it's a circle at all. We could describe the situation by saying that the IIB circle opens up into a nearly flat spatial dimension. This might remind you a little of the duality between Type IIA string theory and M-theory. In that duality, an eleventh dimension opens up when you make string interactions very strong.

I promised to explain the term "deformation," which I used in the previous section in connection with string dualities. Changing the size of a circle is one example of a deformation. Changing the strength of string interactions is another. In general, what I mean by a deformation is any change that takes place smoothly. A string duality is *not* a deformation. Instead, it's a relation between two theories, each of which can be deformed. Or you could think of a string duality as just a change of perspective: you describe the same physics in two different ways. Sometimes one is much simpler than the other: for example, Type IIB string theory is much simpler when interactions are weak than when they are strong. And yet S-duality exchanges weak and strong interactions. This relation of simplicity to complexity is what my analogy to Fred Astaire and the slimy alien was supposed to capture. What this analogy doesn't capture so easily is that you can smoothly change the strength of string interactions, from weak to strong or strong to weak. It's as if we could gradually deform Fred Astaire into the alien, while at the same time the alien turns gradually into Fred. One of the central insights of the second superstring revolution was that by deforming a theory in various ways, and passing through the various known dualities, one can get from any string theory to any other. I've introduced you to three: T-duality, relating Type IIA string theory to Type IIB, S-duality, relating Type IIB to itself, and the duality that relates Type IIA string theory to M-theory. There are three other superstring theories, and dualities that relate them, but I don't think it would help to discuss them here.

On a first pass, I expect it is hard to keep track of all the different branes and dualities. But I hope that one point came across clearly: spatial dimensions in string theory are mutable. They come, they go, they shrink and grow. It's not clear to me that the eventual relation of string theory to the world has to involve extra dimensions per se. If spacetime is only an approximate notion when dimensions are small, maybe the right description of the world involves four big dimensions—the ones we know and love—and then some more abstract mathematical qualities that stand in for extra dimensions. There are constructions like this dating back to the first superstring revolution, but they're not very popular these days.

STRING DUALITIES

Gravity and gauge theory

One particular string duality has become a field unto itself: the gauge/string duality. It is unusual in that it relates type IIB string theory not to another string theory, but to a gauge theory. I discussed gauge symmetry in chapter 5 at some length. Let me recap the essential points. Gauge symmetry guarantees that photons are massless. It guarantees that the axis of the spin of a photon is aligned with the direction of motion of a photon. And it allows us to view electric charge as rotation in an abstract space associated with the gauge symmetry. A gauge theory is any theory whose mathematical description includes a gauge symmetry. Usually that means that the theory includes photons, or things like photons. The theory of light (which is also the theory of electric and magnetic fields) is a simple gauge theory. More complicated gauge theories are of interest not just to string theorists, but also to particle physicists, nuclear physicists, and condensed matter physicists.

You may remember that the gauge symmetry of photons and electrons is secretly the same as the symmetries of a circle. A charged object, like an electron, effectively has some rotation around this circle. We don't have to take this circle as literally as we do the eleventh dimension of M-theory. It's only there in the mathematics to tell us about electric charges and their interactions with photons. One aspect of this mathematics is that photons themselves do not carry electric charge: they only respond to it.

It is natural to ask: If the symmetries of a circle are associated with photons, is there a gauge theory associated with the symmetries of a sphere? It turns out there is such a theory. It has three different kinds of photons, corresponding to the

CHAPTER SIX

three ways of rotating a sphere. (In aviation, those three types of rotations are called pitch, roll, and yaw.) What makes them really different from ordinary photons is that they're charged. You may recall that we had an extended discussion about the cloud of virtual particles that surround an electron or a graviton. Again I will recap the main points. There is a clear distinction between gravity, where gravitons proliferate by responding to one another, and electromagnetism, where photons can only proliferate by splitting into electrons, which produces further photons, and so on. The latter case is by far the more tractable. You can keep track of the whole cascade of virtual particles. Photons and electrons are therefore said to form a renormalizable theory. This theory is called quantum electrodynamics, or QED for short. Gravity, on the other hand, is non-renormalizable. This means that the virtual gravitons cascade out of any sort of mathematical control we know how to impose. Now, what about the gauge theory relating to the symmetries of the sphere? It turns out it's more like QED than like gravity. It's renormalizable.

A cornerstone of our understanding of the physics inside a proton is a gauge theory called quantum chromodynamics, or QCD for short. It is based on a symmetry group with eight different kinds of rotations. As usual, these rotations don't act in our usual four dimensions: they act in some more abstract mathematical space called "color space." QCD is quite similar to the gauge theory based on the symmetries of a sphere. It's just a bit more complicated because there are eight kinds of rotations instead of the pitch, roll, and yaw rotations of a sphere. Each of these eight rotations corresponds to a particle, similar to the photon. Collectively, these eight particles are called gluons. There are also particles like electrons, called quarks. But while electrons can only have negative charge, quarks can have one of three different types of charge. This charge is called color, and color space is the mathematical tool for keeping track of it. A quark's charge can be red, green, or blue. This is only a manner of speaking: there isn't really any relation to the color we see with our eyes. Just as photons respond to the charge of an electron, so gluons respond to the charge of a quark. But gluons are also colored. They respond to each other the way gravitons do. Unlike the uncontrollable cascade of virtual gravitons from other gravitons, the cascade of virtual particles from a quark is something you can keep track of mathematically. So QCD is renormalizable, like QED. The name was chosen partly because QCD greatly resembles QED, and also because "chromodynamics" means "the dynamics of color." Again, this is a notion of color divorced from what you see with your eyes. Color is just a way of visualizing a mathematical abstraction.

Quarks, gluons, and color-that-isn't-color make QCD sound almost as fanciful as string theory. But unlike string theory, it is experimentally very well tested. It is universally accepted as the correct description of physics inside the proton. It has many odd features, the most notable being that you can never directly measure a quark. This is because it dresses itself with gluons and other quarks to such an extent that you never see anything but bound states of quarks and gluons. Protons are such bound states. So are neutrons. Electrons are not. They seem to have nothing to do with quarks. More properly, they are on an equal footing with quarks: separate and equal. One of the big unverified ideas of modern particle physics is that electric charge might be secretly a fourth type of color. I'll discuss related ideas in chapter 7.

The fluctuations of D3-branes are described by gauge theories similar to QCD. I've already discussed fluctuations of

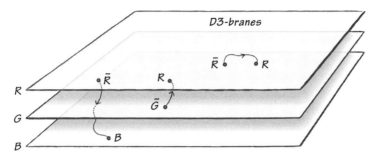

Three D3-branes very close to one another are labeled "red," "green," and "blue." Strings running from one brane to another describe the fluctuations of the branes.

D1-branes. I had two pictures for them: you can think either in terms of a ripple or wave traveling along the D1-brane, or you can think of strings attached to the D1-brane and sliding along it. The latter description generalizes better to D3-branes. Suppose we put three D3-branes on top of one another. Just for the sake of illustration, let's distinguish them by labeling one red, one blue, and one green. If a string runs from the red brane to the blue brane, then intuitively, it's colored. It's purple, right? Well, no—it turns out that's taking the metaphor of color too far. The proper way to describe the string's color is simply that it runs from red to blue. And it turns out that this is exactly the kind of color that gluons possess. You can now almost understand why there are eight types of gluons. There's red-to-red, red-to-blue, red-to-green; three types starting with blue; and three more starting with green. That's nine total. Oops, one too many! Unfortunately, I'd have to introduce an unreasonable amount of extra mathematics in order to explain why one is extra.

Up to this little problem of the extra gluon, we've seen how gluons come out of strings on a trio of D3-branes.

STRING DUALITIES

Quarks are trickier. In the interests of getting to a punch line, I'll leave them out. Clearly, it was just a choice to have three D3-branes together. I could have had just one. Then I would have just photons, like in electromagnetism. I could have had two, and then I'd get the theory I mentioned earlier where the gauge group was the symmetry of a sphere. Or I could have some large number N, in which case there are lots and lots of gluons: about N^2.

The next step is to remember that when many branes come together, the best description of them is in terms of a zero-temperature black hole. I explained this in chapter 5 in the case of D0-branes. The story is similar for D3-branes. With a lot of them on top of one another, they distort spacetime in their vicinity in such a way that there is a black hole horizon. The horizon surrounds the D3-branes in a way that's hard to visualize, because there are too many dimensions. The shape of the horizon is like a cylinder. It's round in some directions and straight in others. But a cylinder is round in one direction and straight in one. It has no additional dimensions beyond those two. The horizon surrounding D3-branes is round in five directions and extended in three. So in total it's eight-dimensional. Tough stuff! And pretty distant from QCD, or so it seems. If there is extra vibrational energy on the D3-branes, then the horizon grows a little and acquires a finite temperature.

An important part of the gauge/string duality is the realization that you can apply formulas like $E = k_B T$ to the vibrations on D3-branes and gain an understanding of the temperature of the horizon surrounding the D3-branes. Let me try to explain why this is now regarded as a string duality. There are two ways to describe D3-branes at finite temperature. One is to keep track of all the open strings sliding around on the D3-branes. Another is to keep track of the horizon surrounding the D3-branes. The two views are complementary in the following sense. If there is a horizon, then you can't say for sure what's inside it. In other words, the existence of a horizon prevents you from keeping track of the strings on the D3-branes. At least, you can't keep track of them one by one. What you can do is to track aggregate quantities such as their total energy. It turns out that when there's a horizon, the gluons are strongly interacting. They split and join frequently. They flicker in and out of existence. They dress themselves with complicated cascades of other gluons. As in strongly interacting string theory, they're scarcely identifiable as gluons. The emergence of the horizon is somewhat like the growth of the extra dimension of M-theory. It explains strong coupling dynamics of gluons in a language that requires extra dimensions.

There's much more to the gauge/string duality than keeping track of the energy of thermal gluons. The right way to understand it is that the curved black hole geometry near the D3-branes is precisely equivalent to the gauge theory of gluons on the D3-branes. This is a strange statement because the curved geometry is ten-dimensional, whereas the gluons only know about four dimensions. It is also strange because it relates a theory with gravity (string theory in the vicinity of the D3-branes) to a theory with no gravity (the gauge theory on the D3-branes). It seems at first more focused, more narrow than other string dualities. T-duality, for example, relates the whole of Type IIB string theory to Type IIA string theory. It includes rules for mapping every type of D-brane to every other type. The gauge/string duality seems restricted to the dynamics of just one type of brane: the D3-brane. But in fact, other branes enter into the gauge/string duality in interesting ways, for example to allow for quarks as well as gluons. I'll have more to say about the gauge/string duality in chapter 8, where I'll describe some attempts to connect it to the physics of heavy ion collisions.

As a closing thought for this chapter, let me point out that string dualities are different from symmetries, even though both express a notion of sameness. The two things related by a string duality can experience different numbers of dimensions. As we just saw, one can include gravity while the other doesn't. That seems really different from a symmetrical object like a square. All its corners are the same, and the symmetries of a square explain precisely how self-similar a square is. On the other hand, there are some string dualities where the two sides seem more nearly mirror images. For example, Type IIA and IIB string theory are really very similar, despite having different types of branes. String dualities show up in low-energy supergravity in a way that's closely connected to ordinary symmetries like the symmetry of a square. It's possible that our understanding of string dualities is incomplete, and that a more unified view of them would make the analogy with ordinary symmetries more precise. There are hints of such a unified view, but too much of what we do understand is restricted to the low-energy theories.

CHAPTER SIX

Chapter S E V E N

SUPERSYMMETRY
AND THE LHC

IN THE SUMMER OF 2008, WHEN THE CONSTRUCTION OF THE Large Hadron Collider, or LHC, was almost complete, I visited the site and took a tour of one of the main LHC experiments. Mostly I was there for a conference, but the tour was really fun. The experiment I visited, called the Compact Muon Solenoid, is about the size of a three-story building. I saw it in the last stages of being put together. A massive cone-shaped endcap was being fitted into the main barrel-shaped body of the detector. Its design is a little like a digital camera, but every part of it looks inward toward its center, where high-energy collisions of proton beams occur.

When the conference finished, I took the opportunity to do a bit of mountaineering in the French Alps. Nothing hard—just a little alpine climbing. The last thing I did was to climb up a ridge to the Aiguille du Midi, from which my climbing partner and I caught a téléphérique that took us back down to the town below. The ridge we climbed is famously narrow, heavily trafficked, and snow-covered. For some reason everyone seems to climb it roped up. I've never quite approved of the practice of climbing roped when no one is tied to a solid anchor. If one person falls, it's hard for the others to avoid being pulled off their feet. Usually I think it's better to trust yourself and climb unroped, or else anchor and belay. But I'll admit that I climbed the ridge roped up to my climbing partner like everyone else. My partner was a very solid climber, and the ridge isn't really that tough.

In retrospect, I think that roped teams climbing a narrow ridge provide a good analogy to the Higgs boson, which is one of the things LHC experimentalists hope to discover. Think of it this way. Standing on top of the ridge, you're balanced precariously. Both sides are quite steep, so if you fall off either way, you're a goner. Tachyons in string theory are like that: they're balanced in an unstable way, and the slightest perturbation sends them sliding down a slope to a fate that string theorists are only starting to understand. But there's more. Let's say you have eight people roped up, and the first one falls off to the left. The second one is probably going to get pulled off to the left too. The third hasn't got a prayer of holding the weight of two falling climbers, so he's going to go too. The really right thing to do in this circumstance is to jump off the *other* side of the ridge and trust the rope. But for some reason this is hard to do.

Back to tachyons and the Higgs boson: The point I want to make is that tachyons usually indicate an instability at every point in space, and that those instabilities are "tied together" like roped-up climbers. If a tachyon starts rolling in one direction at one point in space, it tends to drag the tachyons nearby along with it.

CHAPTER SEVEN

The Higgs boson describes what happens after the tachyons finish "condensing." (Tachyon condensation is the technical term for the roll-off from the ridge.) Let's imagine a merciful outcome for the unlucky climbing team that falls off a high ridge: they slide down to the bottom of a valley and come gently to a halt. Let's suppose they're so tired out that they can't make it back up the slope. Instead they wander around near the bottom, occasionally making small forays up the slope and then sliding back down. This is roughly what the Higgs boson is like. Once the tachyons condense everywhere in spacetime, the quantum fluctuations around their resting point are Higgs bosons.

A problem with the analogy between roped climbers and the Higgs boson is that the direction in which the Higgs boson moves is not one of the familiar three directions of space. Instead, it's like an extra dimension of spacetime—but mathematically tamer. It's also crucial to realize that the Higgs boson is hypothetical. It might simply not be there.

Despite the hypothetical status of the Higgs boson, there's a lovely, deep theory based upon it that has reigned supreme for decades as the best empirical description of particle physics. It is called the Standard Model. "Standard" reminds us that it is widely accepted. "Model" evokes the fact that it is still provisional, and almost certainly incomplete. There's much more to the Standard Model than just the condensation of a tachyon. Among other things, it says that the Higgs boson controls the mass of subatomic particles such as electrons and quarks. It's been hoped for years that an accelerator called the Tevatron, near Chicago, would find the Higgs. And there's still some hope that it might. But the LHC should find either the Higgs boson or else some other weird stuff that stands in its place. An earlier facility in Texas, the Super-conducting Super Collider, had an even greater chance to make dramatic new discoveries. Construction began in 1991. Then, in 1993, Congress pulled the plug. In doing so they probably saved the American taxpayer ten billion dollars. I think it was a bad choice. It certainly means that America has ceded its dominance in experimental particle physics to Europe for the foreseeable future. Fortunately, European nations stayed the course in building the LHC. And Americans have contributed significantly to the LHC effort. So we still have a shot at big, important discoveries.

The weird math of supersymmetry

A great hope for the LHC is that it could discover super-symmetry. This is the symmetry that keeps superstring theory balanced. It does so by excising tachyons, as I described briefly in chapter 4. It is also the symmetry that relates gravitons and photons and guarantees the stability of D0-branes, as I discussed in chapter 5. Supersymmetry and string theory are logically distinct. But they're deeply intertwined. Discovering supersymmetry would mean that string theory is on the right track. There might still be skeptics who would point out that you can have supersymmetry without string theory. While this is true at some level, I think the existence of supersymmetry without string theory would be too great a coincidence to be believed.

But just what is supersymmetry? I've danced around this question a few times already in this book. Let me now try to plow straight into it. Supersymmetry calls for extra dimensions of a very peculiar sort. The dimensions we're used to, and also the extra dimensions of string theory that

I've discussed so far, are measured in length. And length is a number: 2 inches, 10 kilometers, and so forth. You can add two lengths to get another length, and you can multiply two lengths to get an area. The extra dimensions of supersymmetry are not measured in numbers. At least, not ordinary numbers. They are described by anti-commuting numbers, which are the cornerstone of the weird math of supersymmetry. Anti-commuting numbers also play a role in describing electrons, quarks, and neutrinos, which are collectively called fermions. Even though I haven't defined "anti-commuting" or "fermion" yet, I'm going to use these words in the interests of calling things by their true names, or as close as I can come to their true names without using too much math. The extra dimensions of supersymmetry are called fermionic dimensions.

Here's what these funny fermionic dimensions feel like. You can choose to move into them or not, just as you can choose to move forward or sideways. But if you move in a fermionic dimension, there's only one "speed" at which you can move. Speed itself is only a rough analogy to what it means to move in a fermionic dimension. What's closer— though still incomplete—is spin. Moving in a fermionic dimension means that you're spinning more than if you didn't move. The spin of a top can be bigger or smaller according to how hard you twist it before letting it go. But fundamental particles can only have certain amounts of spin. The Higgs boson (if it exists) has no spin. An electron has a minimal amount of spin. A photon has twice as much; but as we learned earlier, the axis of its spin has to align with its motion. A graviton has twice as much spin as a photon. And that's it. No fundamental particle can spin more than a graviton. If supersymmetry is right, a Higgs boson isn't moving at all in the fermionic dimensions. An electron is moving in only one. A photon is moving in two. The story gets a little more vexing for gravitons: depending on how many fermionic dimensions there are, it might be that part of a graviton's spin is due to its motion in the fermionic dimensions, and part of it is intrinsic to the ordinary dimensions of spacetime.

To summarize, there's a sort of exclusiveness about these fermionic dimensions. Either you experience them (like an electron does) or you don't (like a Higgs boson). This exclusiveness has another manifestation, called the exclusion principle. It says that no two fermions can occupy the same quantum state. Electrons are fermions, and there are two of them in a helium atom. These two electrons can't be in the same state. They have to vibrate differently around the helium nucleus, or they have to have different spins—or both. The definition of a fermion is something that obeys the exclusion principle.

Bosons are all the other particles: photons, gravitons, gluons, and the Higgs boson—if it exists. Bosons are very different from fermions. Not only are they allowed to be in the same state as other bosons; they prefer it. Supersymmetry is a relation between bosons and fermions. For every boson, there is a fermion, and vice versa. For example, if the Higgs boson exists and supersymmetry is correct, then there is a Higgs fermion, sometimes called the Higgsino, or occasionally the shiggs. Whatever you call it, the Higgsino would basically be a Higgs boson that's moving in one of the fermionic dimensions.

Fermionic dimensions are hard to draw. The way they're usually studied is through some strange rules of algebra. Let's say there are two fermionic dimensions. You have a letter for

CHAPTER SEVEN

each one: let's use a and b. You can add and multiply them, and most of the usual rules of algebra apply. For example:

$$a + a = 2a$$
$$2(a + b) = 2a + 2b$$
$$a + b = b + a.$$

But there are some very strange rules for multiplying fermionic quantities together:

$$a \times b = -b \times a$$
$$a \times a = 0$$
$$b \times b = 0.$$

The way to think about this is that 1 means you are moving only in bosonic dimensions; a means you're moving in the first fermionic dimension; and b means you're moving in the second fermionic dimension. If you try to move twice as much in the first fermionic dimension, you might try to describe yourself as $a \times a$. The identity $a \times a = 0$ says that the motion you just tried isn't allowed. The meaning of $a \times b = -b \times a$ is harder to explain. To see why it is naturally part of the algebra of fermionic quantities, let me rephrase the rules of multiplication in the following way: $q \times q = 0$ for any combination q of fermionic quantities. If $q = a$, you get $a \times a = 0$. If $q = b$, you get $b \times b = 0$. But what do you get if $q = a + b$? Let's multiply it out:

$$(a + b) \times (a + b) = a \times a + a \times b + b \times a + b \times b.$$

I'll bet you used to do this kind of manipulation in high school math classes. My teachers called it a FOIL expansion. The first term on the right side of the equation is the first term from the left-hand factor times the first term in the right-hand factor. First-first is abbreviated "F." The second

SUPERSYMMETRY AND THE LHC

124

term is the product of the outer terms from the left-hand side: that is, *a* from the first factor and *b* from the second. Outer is abbreviated "O." The third term is the product of the inner terms: *b* from the first factor and a from the second. Inner gives us the "I" in FOIL. The fourth term is the product of the last terms from each factor, so "L" for last-last. Now for the punch line. We've *assumed* that $q \times q = 0$ for any fermionic quantity q, whether it's *a*, or *b*, or $a + b$. If we use this assumption, then the FOIL expansion I just worked through is

$$0 = a \times b + b \times a.$$

That's the same as $a \times b = -b \times a$, which is what I wanted to explain. A key idea to take away from this discussion is that fermionic dimensions require some funny algebra. You might even say that fermionic dimensions are nothing more than the algebraic rules that describe them.

Supersymmetry is symmetry under rotations between bosonic dimensions and fermionic dimensions. What does this mean, exactly? Well, symmetry is a notion of sameness, like how a square looks the same if rotated by 90°. A bosonic dimension is one of the usual ones, like length or width. (The six extra dimensions of string theory are bosonic dimensions too, but that doesn't matter just now.) Fermionic dimensions amount to the funny rules of algebra I explained in the previous paragraph. A rotation between a bosonic dimension and a fermionic dimension means that if a particle was moving in a bosonic dimension before the rotation, then afterwards it isn't; and if it wasn't before, then afterwards it is. Physically, if you start with a boson and rotate it into a fermionic dimension, then it becomes a fermion. If you think of this rotation in terms of mathematics, you would replace the number 1 (representing the bosonic dimension) by *a* or *b* (representing a fermionic dimension). Where the notion of sameness comes in is that the fermion you end up with has the same mass and charge as the boson you started with. This brings us to one of the most distinctive predictions of supersymmetry: it implies that for every boson, there's a fermion with the same mass and charge, and vice versa.

One thing we know for sure is that the world isn't perfectly supersymmetric. If there were a boson with the same mass and charge as an electron, we'd surely know about it. For one thing, it would totally alter the structure of the atom. What might be happening is that supersymmetry is "broken," or violated, by some mechanism similar to tachyon condensation. If the idea of a strange new symmetry that isn't really a symmetry starts making you feel like there's smoke and mirrors somewhere, I don't blame you. Like much of string theory, supersymmetry is a long chain of reasoning without firm contact with experimental physics.

If the weird ideas of supersymmetry and fermionic dimensions are borne out by discoveries at the LHC, it will be a triumph of pure reason beyond anything that's happened in our lifetimes. A lot of people really have their hopes set on it. But wishing won't make it so. Supersymmetry is there, in some approximate form, or it isn't. Frankly, I'll be surprised either way.

The theory of everything—maybe

Here is a synopsis of the canonical ideas about how string theory describes the real world. String theory starts out with ten dimensions. Of course, I'm talking about superstring theory here, so there are some additional fermionic dimensions; but let's put them aside for a moment. Six of the ten dimensions get rolled up in some more or less complicated way. There is a preferred way to do it that exploits the mathematical structure of superstrings, taking advantage of both the supersymmetry and some other properties of the worldsheet description. The rolled up dimensions are small—perhaps a few times bigger than the typical size of a vibrating string. All the overtone modes are so massive that they play essentially no role in physics accessible at the LHC. The most important information comes from the lowest vibrational modes of the strings. In some scenarios, there are D-branes, or other branes, peppered through the extra dimensions, and they introduce additional quantum states for strings that can be relevant to LHC physics.

After rolling up six of the ten dimensions of string theory, what you really want to know is what physics in the remaining four dimensions is like. The answer is that there is always gravity, and there is usually also a gauge theory not unlike QCD. Gravity comes from a massless string state that is quantum mechanically smeared out over the extra six dimensions. The gauge theory can come either from similar smeared-out string states, or from extra string states associated with branes.

Gravity in four dimensions is great—that's what general relativity describes. So the question of whether string theory provides a "theory of everything" mostly comes down to whether the gauge theory you get from rolling up the extra dimensions leads to realistic predictions about subatomic particles. To understand a little more about that gauge theory, first remember that we described the gauge symmetry of QCD in terms of three colors: red, green, and blue. Well, the best-motivated candidates to describe everything— quarks, gluons, electrons, neutrinos, and all the rest—have

CHAPTER SEVEN

at least five colors. String theory constructions can accommodate five-color gauge symmetry in several natural ways. We don't see those five colors yet because something serves to distinguish two of them from the other three. That something could be similar to a Higgs boson, but there are other ideas. To understand what makes five special, remember the enumeration of fermions: there are quarks, electrons, and neutrinos. Quarks come in three colors, but electrons and neutrinos each come in just one. Three plus one plus one is five. It's really that simple.

After the dust settles, the best string constructions yield low-energy physics that is strikingly like what we have already seen in particle physics experiments. Typically, they require supersymmetry and demand not just one Higgs boson, but two; and they require a whole host of other particles whose masses are comparable to the Higgs. They also accommodate a very small mass for the neutrino. And they incorporate gravity as described by general relativity. All in all, this is pretty impressive: certainly, no other theoretical framework for fundamental physics does as well in supplying the right ingredients with the right dynamics. If string theorists could somehow hit upon just the right construction, it would be the "theory of everything": that is, it would include all the fundamental particles, all the interactions they experience, and all the symmetries they obey. Nothing would be left but to solve the equations of this theory and predict every measurable quantity in particle physics, from the mass of the electron to the strength of interactions among gluons.

There are, however, some persistent difficulties. A lot depends on the size and shape of the extra six dimensions. There is no reason that we know of why these dimensions couldn't be flat. In other words, we don't know of any dynamics that would force us to live in four dimensions instead of ten. One possibility is that all dimensions were tightly curled up in the early universe, and that for some reason it is easier for just three of them to unroll into the spatial dimensions of our experience than for all nine to unroll. But this still doesn't explain in any detail why the extra dimensions have the shape they have. To make things worse, the extra dimensions tend to be floppy. To understand what I mean by that, let me remind you of our discussion of the clump of D0-branes. It too had a certain floppiness, in that each D0-brane was only barely restrained from flying away from the others, and D0-branes outside the clump were neither attracted nor repelled by it. The floppiness of extra dimensions means that they could change their size or shape just as easily as a D0-brane could escape from the clump.

A lot of effort has gone into finding ways to tie up these extra dimensions so that they don't flop around anymore. The typical ingredients are branes and magnetic fields. It's easy to understand the role of branes. They're like packing cord you tie around a package. But suppose your package was really mushy. You'd need lots of packing cord to keep the package from bulging out one way or another. The magnetic fields play a similar role in that they stabilize the extra dimensions in some way.

The picture you wind up with is that the extra dimensions are complicated. There are probably many, many ways to tie them up so that they can't flop around. This myriad of possibilities is sometimes regarded as a good thing, because of another problem, called the cosmological constant problem. Briefly, if there is a cosmological constant, then the three dimensions of spacetime themselves have a tendency to bulge out over time. We see from astronomical observations that

CHAPTER SEVEN

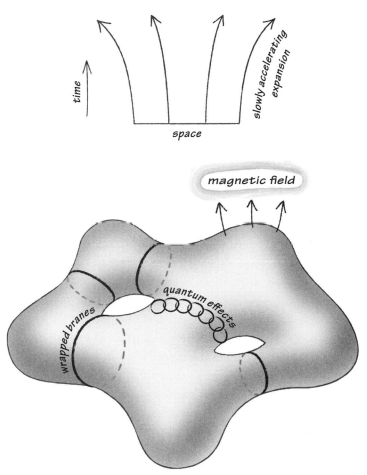

The world according to string theory (maybe). The usual four
dimensions (above) have a slight tendency to expand. The extra six
dimensions (below) have to be tied up with wrapped branes and other
tricks to keep them from bulging out or changing shape.

most galaxies are moving away from us, and this is interpreted as an expansion of space itself. What a cosmological constant would do is to cause that expansion to accelerate. In fact, observations over the last ten years seem to indicate that the expansion of the universe *is* accelerating in a way that's consistent (so far) with a very small cosmological constant. If we want to describe the world using string theory, then it seems we need to tie up the extra six dimensions so that they can't move at all, but leave the usual three with just the slightest tendency to expand and to accelerate their expansion. It's hard to figure out exactly how to do that. But it does seem that the number of ways of tying up the extra dimensions is tremendously large. According to some string theorists, with so many possibilities available, there must be at least a few where everything works out just right, so that the cosmological constant comes out in an acceptably small range. Our universe just happens to be one in which the extra dimensions are tied up in just the right way. If it weren't—the argument goes—intelligent life would probably be impossible, so we wouldn't exist. Turning things around, our existence implies that the universe we inhabit has a small cosmological constant. Altogether, I find myself unconvinced that this line of argument is useful in string theory.

String theorists have been hammering on the question of how to cook up the theory of everything for more than twenty years. Rolled up extra dimensions always play a role. The more we learn about string theory, the more possibilities there seem to be. It's embarrassing. Perhaps it is worth comparing the difficulty of getting fully realistic four-dimensional physics out of string theory to a long-standing problem in another corner of theoretical physics: high-temperature superconductivity. Discovered starting in 1986, high-temperature

CHAPTER SEVEN

superconductors conduct large amounts of electricity without significant energy loss. High-temperature is perhaps an exaggeration: the temperatures in question are comparable to the freezing point of air. But that's a lot hotter than previously known superconductors, and there are already some important industrial applications. Theoretically, though, it's very hard to understand how high-temperature superconductivity works. There's a theory from the 1950s that explains ordinary superconductors, and it's based on tying pairs of electrons together. The force that pairs them together is based on sound. The electrons sort of "listen" to one another, over distances many times the size of an atom, and then coordinate their motion so as to avoid energy loss. Magical. But also fragile. Too much thermal motion prevents this pairing from occurring: it's as if the electrons can't "hear" one another over the din of thermal noise. It's believed that no amount of tinkering with this 1950s explanation, in which electrons coordinate their motions through sound waves, will account for the remarkable properties of high-temperature superconductors. Electrons probably still pair up in these materials, but over a much shorter distance, in a much stronger way. They seem to take advantage of fine-grained features of their environment to pair up. There are some compelling theoretical ideas about how this happens, but I do not think the problem is solved.

Solved or not, high-temperature superconductivity might offer some lessons for string theory. The main one is that pure reason is often not enough. High-temperature superconductors were an experimental discovery, and theory has been struggling ever since to catch up. The correct theory of the world could be quite different from what we are now capable of imagining. The fragile pairing of electrons through sound waves reminds me of the floppiness of extra dimensions: just barely holding together. It could be that the way string theory really relates to the world is as different from these tied-up bundles of branes, magnetic fields, and extra dimensions as the modern explanations of superconductivity are from theories of the 1950s. And it might take at least as long to figure it all out.

Particles, particles, particles

In chapter 5, I alluded briefly to the long list of known elementary particles: photons, gravitons, electrons, quarks (six kinds!), gluons, neutrinos, and a few others. Explaining the whole list wouldn't add much to the obvious point that this is a lot of different particles, each with its own peculiar properties and interactions. Long lists of objects cry out for a unifying theory with fewer elementary objects and a deeper level of explanatory power. Chemistry's periodic table receives such a unifying treatment through atomic theory. Helium, argon, potassium, and copper are all as different as they ever were in chemical reactions. But atomic theory reveals that they are all composed of electrons in quantum states of vibration around an atomic nucleus composed of protons and neutrons. The long list of elementary particles might receive a unifying treatment in terms of string theory. As for the long list of objects in string theory—D-branes, solitonic 5-branes, M-branes, and so on—no one knows how or whether they might be unified beyond the level of string dualities.

The most massive particle discovered to date is the top quark. Its mass is about 182 times the mass of a proton. It was discovered in 1995 by large experimental collaborations at the Tevatron, which is the premier particle accelerator in the United States. Protons and anti-protons are whirled

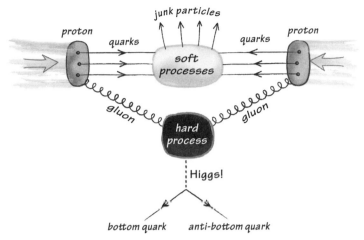

A proton-proton collision at the LHC might produce a Higgs boson in the manner indicated here. In the process I have drawn, the Higgs promptly decays into a bottom quark and an anti-bottom quark, which can be detected. But the "junk particles" can cause confusion as to what really happened.

around a large ring (about 6.3 km around) and smashed together in head-on collisions. When they hit, each of them has an energy 1000 times as large as its rest mass. It's not surprising that such collisions can produce a top quark: there's plenty of energy available. In fact, there's apparently enough energy to make a particle ten times as massive as the top quark: $1000 + 1000 = 2000$ proton masses. Unfortunately, it's all but impossible for all this energy to go into a single particle. This is because protons and anti-protons have structure. Each contains three quarks and also some gluons. When the proton and anti-proton collide, most of the quarks and gluons miss one another, or experience only glancing collisions. What's interesting is the situation where one quark or gluon from the proton hits hard against one

SUPERSYMMETRY AND THE LHC

from the anti-proton. Such a hard hit—more commonly described as a "hard process"—is what creates top quarks at the Tevatron. Hard processes should also create Higgs bosons, if they exist. Because hard processes involve only one quark or gluon from the proton and one from the anti-proton, the energy available to make top quarks is only a fraction of the total energy of the collision.

The LHC will collide pairs of protons with a total energy of about 15,000 times the mass of a proton. The amount of energy available in a hard process might be about a tenth of this—sometimes more, sometimes less. Speaking in round numbers, the LHC can be expected to produce particles copiously whose rest mass is up to 1000 times the mass of a proton. Heavier particles should also be produced, up to perhaps 2000 times the mass of a proton.

But the heavier a particle is, the rarer it will be that a hard process has enough energy to produce it.

Just what kinds of particles should we expect the LHC to discover? At time of writing, the honest answer is: We're not sure, but there had better be something. I do not mean this in the sense that the LHC will be a big waste of money if it doesn't discover anything—though that's obviously true. What I mean is that there's a good argument, independent of ideas of supersymmetry or string theory, that there is something lurking at the energy range that the LHC will explore. It could be just the Higgs boson. Most likely, it's the Higgs boson and some other particles. If we're lucky, it will be supersymmetry. The argument that something has to be there relies on renormalization. I gave a brief, qualitative account of renormalization in chapter 4, but to remind you, it is the mathematical machinery that allows us to keep track of the cloud of virtual particles that surround an electron,

CHAPTER SEVEN

or indeed any particle. This machinery only works if there's something like a Higgs boson in a range of energies that the LHC should explore. For it to work smoothly, there has to be something like supersymmetry in addition to the Higgs. But let's not forget that our mathematical machinery is not the world. We could just be wrong. There could be something at the LHC that we haven't imagined. That would be the most exciting possibility of all. Or—despite all our well-reasoned expectations—there could be nothing to see.

Let's get back to supersymmetry, which is a favored candidate to describe LHC physics. As I explained earlier, a striking prediction of supersymmetry is that for every particle that we know, there's a new one with the same mass and charge and essentially the same interactions, but different spin. We know the electron. Supersymmetry predicts a super-electron, or "selectron" for short. We know the photon. Supersymmetry predicts a super-photon, more commonly known as a "photino." Likewise, supersymmetry predicts squarks, gluinos, sneutrinos, and gravitinos. Even the Higgs would get a super-partner, usually called the higgsino (but occasionally the shiggs). As I also explained earlier, supersymmetry can't be exactly right: for example, we know there isn't a selectron with the same mass as the electron. Approximate or "broken" supersymmetry still predicts that there are selectrons, photinos, sneutrinos, and all the rest. But their masses could be quite a bit larger than the particles we have discovered so far. It is reasonable to assume that most or all of these super-particles (and yes, we call them sparticles) have masses within reach of the LHC. If that's true, then the LHC could be the most prolific discovery machine in history, turning up not just a handful of new fundamental particles, but a dozen or more.

SUPERSYMMETRY AND THE LHC

A symmetry that demands a collection of new particles equal in size to everything presently known might seem like a step backward rather than forward. After all, aren't we supposed to reach for unifying pictures with more explanatory power in terms of fewer ingredients? That's exactly how I felt about supersymmetry when I first learned about it. But here's a comparison worth pondering. The equation for an electron, discovered in the 1920s, led to a very unexpected prediction: the existence of an anti-electron, more commonly called a positron. Soon, physicists were predicting an anti-particle for almost every particle they knew. And they found them! For me, supersymmetry doesn't have the same aura of inevitability. It's not needed to describe particles we do know, in the way that the equation for the electron was needed. But perhaps it's not fair to compare foresight and hindsight.

It's one thing for a particle to exist with a mass in the right range for the LHC to find it, and quite another to actually discover it. That's because it's complicated to sort through all the junk that comes out of a collision and reconstruct what happened. It's actually possible that the Tevatron has been producing Higgs bosons for years, but reconstructing them requires such subtlety that they have escaped notice. In fact, physicists generally favor a mass range for the Higgs boson of no more than 150 proton masses: lighter than the top quark! Sparticles might actually be easier to find at the LHC than the Higgs. Gluinos in particular should be produced copiously, if they are in an accessible mass range. Equally important, they are predicted in many supersymmetric theories to go through a spectacular chain of decays that should be relatively easy to pick out of the data. In this chain of decays, the gluino sheds some of its rest energy by

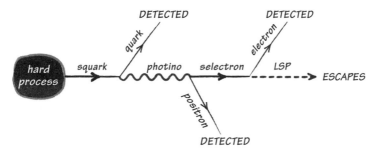

Decay of a squark into several detected particles and an LSP, which escapes undetected.

turning into a different type of sparticle. Then that new sparticle sheds some of its rest energy in the same way. After several such steps, one is left with the lightest sparticle. The lightest sparticle is often abbreviated LSP. It's usually assumed that the LSP won't decay at all, but instead will escape undetected. If all this is right, then what detectors at the LHC will observe is not the superpartners, but the particles they shed during their decays to the LSP.

Before telling you more about the LSP, I should mention one of the unfortunate facts about the LHC: even if it discovers things that look like sparticles, it will be tricky to say for sure whether they are unambiguous evidence of supersymmetry. This is basically because proton–proton collisions are messy. Lots of particles come out. Known interactions among quarks and gluons are so strong that they can mask new phenomena. And it's hard to determine the spin of a newly discovered particle. For all these reasons, physicists have advocated building a companion machine to the LHC, called the International Linear Collider, or ILC. It would collide electrons and positrons. Such collisions provide a much cleaner experimental environment. It would be possible to distinguish more cleanly than at the LHC between supersymmetry and alternative theories. But the ILC is still only a proposal. The dark fate of the Superconducting Super Collider shows how hard it is to bring such proposals to fruition.

Let's get back to supersymmetry. The LSP, if it exists, could be the most important discovery of all, because it could be the dark matter that draws galaxies together. For decades, cosmologists and astronomers have puzzled over the total mass of galaxies. They can count the stars in a galaxy (at least roughly). From that count, they can estimate how much ordinary matter exists in a galaxy. By ordinary matter, I mostly mean protons and neutrons, because they are the main carriers of mass. The problem is that galaxies never seem to have enough mass in ordinary matter to hold together in the way that they do. Hence the hypothesis of "dark matter": there's extra stuff that we don't see in galaxies that was primarily responsible for drawing them together in the first place. Based on a variety of measurements, many or most cosmologists believe that there is five or six times as much dark matter in the universe as there is ordinary matter. But what is it? Various proposals have been floated, ranging from burnt-out stars to subatomic particles. LSPs as dark matter have two main virtues. First, in many of the most realistic supersymmetric theories, they are very massive (more than 100 times the mass of a proton), electrically neutral, and stable—meaning that they never decay into other particles. Second, it is easy to understand how they could have been produced in the early universe in approximately the right abundance—meaning that they comprise five or six times as much total mass today as ordinary matter does.

CHAPTER SEVEN

Altogether, supersymmetry is a wonderful theoretical framework. It's well motivated by strange mathematics. It's beautifully consistent with established particle theory, including renormalization. And it predicts a lot of new particles that we can hope to see at the LHC. Finally, supersymmetry and string theory are so deeply intertwined that it is hard for me to believe one could have supersymmetry in the world unless string theory in some form is correct. Let me put it this way: Supersymmetry is a little like a string duality. It relates particles to sparticles, just as S-duality relates strings to D-branes. Like a string duality, it leaves you wanting more. Isn't there some unifying picture underlying all the particles and sparticles? Shouldn't supersymmetry itself be a hint to what that underlying picture should be? String theory provides a clean answer, where supersymmetry is built in from the start, and where all the particles we know or will discover have a more-or-less unified origin in terms of string dynamics and extra dimensions.

SUPERSYMMETRY AND THE LHC

Chapter E I G H T

HEAVY IONS AND THE
FIFTH DIMENSION

A STRANGE FACT ABOUT THE RELATION BETWEEN SUPERSYM-
metry and LHC physics is that the main ingredients were approximately in place twenty years ago or more. There have
certainly been advances in the past two decades, both theoretical and experimental. The top quark was a big discovery,
although it was long anticipated. The non-discovery of the
Higgs constrains models of supersymmetry in interesting
ways. Theoretical understanding of supersymmetry has
deepened considerably, and the range of possible manifestations of supersymmetry at the LHC has been explored much
better than it was in the late 1980s. But these advances have
been in some sense incremental. Especially now, with the
LHC about to start generating data, one has the sense of
the whole field holding its breath. But it's been holding its
breath a little too long. Supersymmetry is so entrancing that
it has survived literally decades of non-discovery without
losing its place as the main hope. Alternative theories tend

to get calibrated against supersymmetry to such an extent that they start resembling supersymmetry.

Recently, a wholly different route to connecting string theory to the real world has been developed. On the string theory side, it is based on the gauge/string duality, which I introduced in chapter 6. On the real world side, it relates to heavy ion collisions, which I'll describe more in the next section. In such collisions, temperature and density rise so high that protons and neutrons melt into a fluid called the quark-gluon plasma, or QGP for short. There are ways of understanding this melted fluid that have nothing to do with string theory. The right way to characterize the aim of the field is to make string theory one of several quantitatively useful tools for describing the quark-gluon plasma.

This is clearly a less lofty aim than to produce a theory of everything and reveal the ultimate structure of the physical universe. But, at present, the putative connection of string theory to heavy ion physics has two charming features that are missing in the theory-of-everything side of string theory. First, the intellectual content on the string theory side is firmly rooted in string dynamics and the gauge/string duality. This is a more direct access to string theory itself than most theory-of-everything scenarios are likely to offer. That's because connections between string theory and LHC physics are mostly mediated through supersymmetry and the low-energy limit of string theory, where all but the lightest string states drop out of the physics. Second, string theory calculations have already been compared to experimental data, with some degree of success. Caution is still very much in order, and there are significant critiques and disagreements about how and whether string theory relates to heavy ion collisions. Nevertheless, this field is producing

HEAVY IONS AND THE FIFTH DIMENSION

the closest contact to date between modern string theory and experimental physics.

The hottest stuff on Earth

The Relativistic Heavy Ion Collider (RHIC) is a particle accelerator located on Long Island, not far from New York City. Its basic design is similar to the Tevatron and the LHC. But it is relatively puny: it can only accelerate subatomic particles to an energy of about 100 times their rest mass. The Tevatron reaches a factor of 1000, and the LHC will reach as high as a factor of 7000 for protons. The big difference between the Tevatron and RHIC is that RHIC accelerates gold nuclei. There are almost 200 nucleons in a gold nucleus. (Remember, a nucleon is a proton or a neutron.) Gold was chosen because it has a big nucleus, and for some technical reasons relating to how you start accelerating it. When the LHC collides heavy ions, the plan is to use lead, which has even a slightly bigger nucleus. There's nothing really special about gold from the standpoint of heavy ion collisions. I'll keep talking about it just because it's the substance of choice at RHIC.

Particle physicists have long been willing to smash anything into anything if it promises to teach them something. But past preference has leaned toward electrons and positrons. There's a good reason for this: electrons and positrons are small and simple compared to atomic nuclei. There is no evidence that an electron is anything but a point particle. Positrons are just like electrons, only with positive charge. Protons are already much more complicated. They contain at least three quarks, and probably some gluons. Collectively, these constituents of the proton (or, equally of the neutron) are called partons:

each one is "part" of a proton. But a proton is more than the sum of its partons. The strong interactions among quarks and gluons inside the proton are like the cascade of virtual particles we discussed in connection with renormalizability. Let me remind you how that went. A quark can emit a gluon. It does so in the way electrons emit photons. A gluon is like a photon, but not entirely. The big difference is that gluons can split into other gluons. They can also split into quarks, or join with other gluons. All this emission, splitting, and joining is the cascade. The particles are called "virtual" because everything is happening inside the proton. You never actually see a single quark or a single gluon on its own: they're always part of a proton, or a neutron, or some other subatomic particle. Physicists describe this by saying that quarks and gluons are confined. They cascade in and out of existence, always inside the confines of the proton.

When you collide protons, one way to think of what happens is that each one interrupts the other in the middle of its cascade of quarks and gluons. One thing that can happen is that a pair of quarks hits very hard. That is the sort of event on which the hopes of the LHC hang: a hard process. A lot of what happens, though, is that quarks and gluons interact more softly. "Softly" here is a relative term. The colliding protons are utterly destroyed when they hit. Fifty or more particles come out of the collision, most of them unstable.

To get a feel for these collisions, think of a car crash where two cars collide head on. In order to avoid contemplating sad and scary things, let's suppose there aren't any passengers, just crash test dummies. I'm thinking of the cars as analogous to colliding protons, and the dummies as analogous to the quarks inside each proton. In favorable circumstances, the dummies are only lightly damaged even when the cars are totaled. That would be like saying that quarks in one proton interact only softly with partons in the other one. In unfavorable cases, a dummy might be badly mangled by a part of the car going the wrong way. That's like a hard collision. Proton-proton collisions are typically a hybrid, with some relatively hard process plus a lot of soft junk happening around it.

Let me hasten to add that there's nothing harmful about occasional high-energy collisions of subatomic particles. Actually, they're happening constantly in the Earth's atmosphere, as highly energetic particles rain down on us and hit some nucleon or other in the air. What goes on at the Tevatron, and what will go on at the LHC, is just a controlled version of something that's been happening literally since the world began. Because so many collisions happen in the same place at particle accelerators, the environment of the collisions is sealed off underground. There would be a lot of radiation hazard to a person down there. But by comparison to nuclear reactors or atomic weapons, the hazard is relatively mild.

Collisions of gold nuclei are at first blush quite similar to proton-proton collisions. Each nucleus is a big blob of nucleons, each one undergoing its internal cascade of partons. During the collision, some partons may hit each other quite hard, while the majority nudge each other more gently. As with proton-proton collisions, the gold nuclei are utterly destroyed. Literally thousands of particles pour forth from a single collision of gold nuclei.

There is something qualitatively more catastrophic about collisions of gold nuclei than of protons. To describe it, let me return, a little wincingly, to the car crash analogy. One of the worst things that can happen is for one or both gas tanks to ignite and explode. Car makers take considerable pains to prevent this, for example by positioning the tank where it is

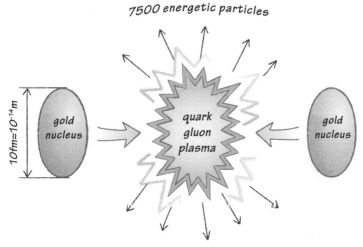

An ultra-high-speed collision of gold nuclei creates a quark-gluon plasma, which decays into thousands of energetic particles.

least likely to be punctured. What happens in a gold-gold collision is a little like the gas tanks exploding shortly after a car collision. There is literally a thermal ball of nuclear fire that forms and then blows itself apart. This ball is hotter than anything you can imagine. A gas tank explosion might reach 2000 Kelvin. The center of the sun is about 16 million Kelvin. Similar temperatures can be reached by detonating a thermonuclear bomb (an H-bomb). Pretty hot, eh? Well, get this: The temperature believed to be attained at RHIC is more than 200,000 times hotter than the center of the sun. That really takes some thinking about. It's way more than white hot: white hot is just thousands or tens of thousands of Kelvin. It's blindingly, radioactively, primordially hot. Protons and neutrons melt in this heat, releasing the quarks and gluons inside them. They form the quark-gluon plasma, or QGP, which I mentioned earlier in this chapter.

HEAVY IONS AND THE FIFTH DIMENSION

In proton-proton collisions, the hard processes that LHC physicists will sift for signs of the Higgs boson and supersymmetry are obscured by all the soft junk that's happening during the same collision. But only a little. When two quarks really hit hard, they tend to bounce off in entirely new directions, and they head out to the surrounding particle detectors pretty much unimpeded by the rest of the proton. In heavy ion collisions, it's exactly the opposite: hard processes happen, but most of the time the resulting particles seem to get "stuck" in the quark-gluon plasma. The extent to which this happens is one of the key properties of the quark-gluon plasma. Bullets shot into water provide a reasonable analogy. You've probably seen movies where James Bond or some similar character is dodging bullets while underwater. The bullets are pictured as whizzing around him, and you see these long bubbly tracks illuminated in funny ways. Well, the reality is that a bullet will penetrate only a few feet into the water. In physicists' terminology, bullets in water have a stopping length of a few feet. One of the distinctive properties of the quark-gluon plasma is that it has a very short stopping length for the particles coming from hard processes: only a few times the size of a proton.

A second distinctive property of the quark-gluon plasma is its viscosity. Considering the tremendously high density of the QGP, its viscosity is surprisingly small. It takes some explaining to understand what this means. On one hand, I think viscosity is a familiar concept to anyone who cooks: honey and molasses are viscous, water and canola oil are much less so. But the contrast one really wants to draw in heavy ion physics is between nearly free-streaming particles, which are deemed highly viscous, and a strongly interacting plasma, which is not. That may seem backwards. Nothing could be less viscous than free-streaming particles, right? If no particle hits another, there's no viscosity, right? Unfortunately, this is completely wrong. Something with truly small viscosity can form flowing layers that slip over one another. Water flowing over rocks does this: a layer of water very close to the rock moves slowly, but layers above it tumble quickly over the stones, lubricated in some sense by the lowermost layer. What if we replaced water with steam but left the rocks where they were? Let's say the steam is constrained to follow the streambed: perhaps we put a cover over the stream to trap the steam. Now, steam is a bunch of individual water molecules that rarely bump into one another. But they do bump into the rocks. Unlike water, steam doesn't form layers that slip easily over one another. It's actually harder to get a given mass of steam to flow through a rough channel than to get the same mass of water to flow through it, because the water is self-lubricating. That's what it means to say that water has a lower viscosity than steam.

Heavy ion collisions create conditions a little like the rocky streambed, except without the rocks or the stream. (Analogies have their limits!) What I mean is that you can tell the difference in heavy ion collisions between a waterlike substance that's low-viscosity—in the sense of being able to flow freely in slippery layers—and a steamlike substance that's basically just a bunch of particles that rarely hit one another. Surprisingly, the best understanding of the data comes from assuming very low-viscosity behavior. Tentative theoretical estimates of the viscosity, based on quantum chromodynamics, were generally well off the mark, predicting that the quarks and gluons would behave less like water and more like steam than they actually do.

The world of heavy ion physics was shaken when it was discovered that black hole horizons have a viscosity that is comparable to the small values one needs in order to understand heavy ion data. This discovery was made in the framework of the gauge/string duality, which I introduced in chapter 6. Subsequent developments seem to indicate that many aspects of heavy ion collisions have close analogies in gravitational systems. The gravitational systems in question always involve an extra dimension. It's not like the extra dimensions of string theory in its theory-of-everything guise. This extra dimension—what I referred to as the fifth dimension in the title of this chapter—is not rolled up. It's at right angles to our usual ones, and we can't move into it in the usual way. What it describes is energy scale—meaning the characteristic energy of a physical process. By combining the fifth dimension with the ones we know and love, you get a curved five-dimensional spacetime. This spacetime encodes temperature, energy loss, and viscosity in geometrical ways. Much effort in the past several years has gone into deciphering how detailed a correspondence can be made between five-dimensional geometries and the physics of the quark-gluon plasma.

In summary: The soft interactions that LHC physicists wish weren't there in proton–proton collisions are multiplied many times over in heavy ion collisions. They lead to the creation of the quark–gluon plasma. The QGP can't be described very well in terms of individual particles. Its properties may in some ways be better understood in terms of black holes in five dimensions, according to the gauge/string duality.

Black holes in the fifth dimension

In chapter 6, I gave a brief introduction to the gauge/string duality. Let me recapitulate a couple of the main points.

CHAPTER EIGHT

A gauge theory similar to quantum chromodynamics describes how strings attached to D3-branes interact. Their interactions can be made stronger or weaker by changing a parameter of that gauge theory. If you make the interactions very strong, then thermal states are best described in terms of a black hole horizon that surrounds the D3-branes. This horizon is hard to visualize, because it is an eight-dimensional hypersurface in a ten-dimensional ambient geometry. A simplification that helps me is to think of the horizon as a flat three-dimensional surface that is parallel to the world we live in, but separated from it in the fifth dimension by a distance related to the temperature. The larger the temperature of this three-dimensional surface, the smaller the separation. This is an imperfect visualization. What it leaves out is that the fifth dimension is not like our usual four. Four-dimensional experience is like a shadow of five-dimensional "reality." But unlike the shadows you see on a sunny day, four-dimensional experience carries no less information than the five-dimensional "reality" behind it. Four-dimensional and five-dimensional descriptions are really equivalent. The equivalence is subtle but precise. It's a metaphor on steroids: every statement you can meaningfully make about four-dimensional physics has a five-dimensional counterpart, and vice versa—at least in principle.

Other string dualities have a similar metaphorical quality. For example, if you recall, the duality between ten-dimensional string theory and eleven-dimensional M-theory includes an equivalence between D0-branes and particles moving around a circle. The special charm of the gauge/string duality is that instead of relating one abstract theory to another, in dimensions beyond anyone's capacity to visualize, one is dealing directly with four-dimensional physics similar to what we

HEAVY IONS AND THE FIFTH DIMENSION

know must describe quarks and gluons. So the equivalent objects on the five-dimensional side of the duality assume special significance. Most importantly for the current discussion, the quark-gluon plasma created in a heavy ion collision relates to a black hole horizon in five dimensions. What really makes this analogy work is that heavy ion collisions produce high enough temperatures to melt the nucleons into their constituent quarks and gluons. Nucleons themselves are relatively difficult to translate into five-dimensional constructions. Individual quarks and gluons are also difficult. But the collective behavior of a strongly interacting thermal swarm of quarks and gluons is easy to translate: the swarm becomes the horizon.

There is undeniably an elusive quality to the gauge/string duality. As well established as it is on technical grounds, it is just strange to have a fifth dimension that isn't really a dimension like the ones we know and love. It's there not so much as a physical direction, but as a concept that describes aspects of the physics of four dimensions. Ultimately, I'm not convinced that the six extra dimensions of string theory as a theory of everything will be more tangible than the fifth dimension of the gauge/string duality.

An additional irony is that the temperature of the black hole is supposed to be enormous, in sharp contrast to the temperature of black holes that might be at the cores of galaxies. Recall our rough estimate from chapter 3: a black hole at the core of a galaxy might have a temperature of a hundredth of a trillionth of a Kelvin. The temperature of a black hole in five dimensions dual to the quark-gluon plasma is more like three trillion Kelvin. What makes the difference is the curved shape of the five-dimensional geometry.

If we accept the picture of the thermal swarm of quarks and gluons as a horizon in five dimensions, then what? Well,

CHAPTER EIGHT

there are lots of things you can do, because the gauge/string duality is a computational bonanza. One of the favorite computations is the viscosity: as computed from the black hole geometry, shear viscosity is very small compared to the density of the plasma, and this seems to match well with a widely accepted interpretation of the data. Some other computations have to do with energetic particles, which (as I described earlier) cannot penetrate very far through the plasma. This phenomenon has an obvious affinity with black hole physics: nothing can get out of a black hole. But that's not quite the same thing as saying that nothing can get very far through a thermal medium. How should the translation go?

There's actually some dispute about the correct answer at the time I am writing this book. I'll explain only one side of the story, and I'll hint a little bit at what the dispute is about.

The side of the story I'll explain hinges on the idea of the "QCD string." This is such an important and well-accepted concept that I'll back off a little bit and explain where it comes from. First, let me remind you that electrons produce a cloud of virtual photons. These photons can be described in terms of an electric field. Actually, any charged object produces an electric field. For example, a proton does. The electric field surrounding a proton tells other protons which way to move in response to the first one. Protons repel one another electrically. The electric field represents that by pointing outward, all around the proton. Protons attract electrons, and this is described by the same electric field: it's just that electrons, being negatively charged, perceive a given electric field in the opposite way that protons do.

Quarks are profoundly similar to electrons, yet also profoundly different. They produce a cloud of virtual gluons that can be understood as a "chromo-electric field," which

152

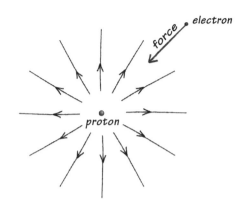

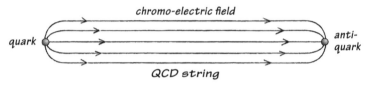

Top: the electric field of a proton points radially outward. Bottom: the chromo-electric field generated by a quark forms into a QCD string that can end on an anti-quark.

tells other quarks which way to move. So far, this is very similar to electrons. But the virtual gluons interact strongly with one another, which is wholly unlike photons. Because of these interactions, the chromo-electric field channels itself into a narrow string—the QCD string—which stretches from one quark to another. There are objects called mesons that are understood to be well described in these terms: two quarks linked by a QCD string. By studying the properties of mesons, you can infer some of the dynamics of the QCD string, and it is in some ways quite similar to the strings of string theory. In fact, such studies are older than QCD or

CHAPTER EIGHT

string theory! They provided the first fodder for speculation that strings could describe aspects of subatomic physics. The modern incarnation of those speculations is one aspect of the gauge/string duality and its relation to QCD. What's really different between modern string theory and QCD is that the string is regarded as the fundamental object, whereas the QCD string is a collective effect of many virtual gluons. However, a lesson of string dualities is not to hold too rigidly to one theoretical construction as fundamental and another as derived: as circumstances vary, so too may the most convenient language to describe reality.

Now imagine a quark that was produced in a hard process and is plowing through the quark-gluon plasma, like a bullet plowing through water. The ideas behind the QCD string should still have some currency: the quark surrounds itself with virtual gluons, these gluons interact among themselves, and some collective tendency toward forming a QCD string might arise. But there's something else going on: all the quarks and gluons in the thermal swarm are interacting with the original quark, as well as with any virtual gluons it might produce. This thermal swarm prevents the QCD string from fully forming. Altogether, the picture might look a little bit like a tadpole: the original quark is like the head, and its attempts to make a QCD string are the tail. The way the tail wiggles and flails through the water is like how the thermal swarm interacts with the virtual gluons. This picture has not been made precise or quantitative (as far as I know) in QCD itself. But there is something similar to it in the gauge/string duality. A string dangles down from a quark into the black hole horizon. As the quark moves forward, the string is pulled forward. But the end that it trails into the black hole horizon is in some

154

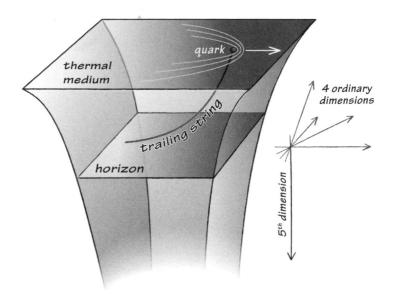

A quark moving through a thermal medium like the quark-gluon plasma trails a string down into the fifth dimension, where it eventually crosses through a black hole horizon. When the quark moves, the string trails behind it and produces a drag force on the quark.

sense stuck. The string pulls back on the quark because it can't get its other end free of the black hole. Eventually, the quark either gives up and stops moving, or falls into the black hole itself. Either way, it doesn't get very far.

The picture I just described is supposed to work best for heavy quarks. Examples of heavy quarks are the charm quark, with a mass about 50% more than a proton, and a bottom quark, with a mass more than four times as much as a proton. Such quarks are almost entirely absent from normal matter, but they are produced in heavy ion collisions. The "ordinary quarks" in normal matter, together with anti-quarks of the

CHAPTER EIGHT

same masses, are produced in heavy ion collisions far more abundantly than heavy quarks. There are attempts to extend the picture of a quark trailing a string to the case of ordinary quarks, but they are somewhat tentative so far.

The bottom line is that the gauge/string duality provides an estimate of how far a heavy quark propagates through a thermal medium similar to the quark-gluon plasma. With such an estimate in hand, the next task is to find out whether it agrees with data.

There are two reasons why this is tricky. First, experimentalists can't train a microscope on the quark-gluon plasma and watch as a heavy quark trundles along and then stops. Instead, their little ball of plasma, including the heavy quark, completely blows itself apart in a time comparable to the time it takes light to traverse a gold nucleus. That's a very, very short time: about 4×10^{-23} seconds, which is forty trillionths of a trillionth of a second. The only things they get to look at are the thousands of particles that come out. It's pretty impressive that they can infer how the charm quark interacted with the medium from inspecting the debris. I think experimentalists would caution that such inferences have to be taken with a grain of salt. They can be 99.99% confident of their measurements, and yet be considerably less certain about how far the average charm quark gets through the plasma.

The second reason why it is tricky to compare a prediction of the gauge/string duality with data is that the string theory computations apply to a theory that is only *similar* to QCD, not to QCD itself. The theorist has to make some translation between one and the other before he or she has a definite prediction to give an experimentalist. In other words, there's some fudge. The best attempts to handle this translation honestly lead to predictions for the charm quark's

HEAVY IONS AND THE FIFTH DIMENSION

156

stopping distance that are either in approximate agreement with data, or perhaps as much as a factor of 2 smaller. A similar comparison can be made for viscosity, and the upshot is that the gauge/string duality produces a result that is either in approximate agreement with data, or perhaps a factor of 2 away from agreement.

All this doesn't sound like a reason to break out the champagne. And yet, agreement between modern string theory and modern experiment to within a factor of 2 is a tremendous novelty for high-energy physics. Fifteen years ago, all the string theorists were toiling away in extra dimensions, and all the heavy ion experimentalists were busy building their big detectors. None of us were even able to imagine the kind of calculations I've described. Now we study each other's papers, we go to the same conferences, we worry about factors of 2, and we try to figure out what to do next. That's progress.

Earlier, I mentioned a dispute about how to translate the stopping of an energetic quark into a process involving strings and black holes. This controversy is not about factors of 2 here or there. Instead, it's about the physical picture one should have in mind in describing the energetic quark. The picture I've described involves a string trailing out from the quark, down into the fifth dimension, and into the black hole horizon. The competing picture is more abstract, but it essentially relies on a U-shaped string configuration where the bottom of the U just grazes the horizon. For lack of better terminology, I'll refer to the two pictures as the trailing string and the U-shaped string. A virtue of the latter is that it purports to describe ordinary quarks. That's good because they're far more abundant, hence easier to study. The U-shaped string leads to predictions about quark energy loss that are once again either on target or within about a factor of 2. The trouble is that the "fudge factors" for U-shaped strings and for trailing strings tend to get chosen differently. Furthermore, proponents of each picture have made specific criticisms of the other. This is not an easy debate to settle. The questions are abstract, the hypotheses are a little different, and the agreement with data is only expected to be approximate. Still, I would take it as a sign of new-found health that string theorists are debating the relative merits of calculations that can be compared at least approximately with data.

What is the future? For heavy ion collisions, I think the answer is that more is better. The more calculations that string theorists can perform, the more handles they should have on the difficult problem of translation. The aim is a reasonably coherent and consistent mapping between five-dimensional constructions and experimentally measurable quantities. It could be that this program hits a roadblock at some point: there could just be some insuperable difference between the string theory constructions and real world QCD. So far that hasn't happened. It could also happen that the string theory calculations peter out because of insufficient ability to handle the technical difficulties. String theory does seem to go in fits and starts: a lot of progress, then relative stagnation, then more progress.

Experiments at the LHC will include slamming lead nuclei together at substantially higher energies than RHIC can reach. (Remember, for purposes of heavy ion collisions, lead and gold are nearly identical.) The data from these collisions should provide a large new stimulus to theoretical approaches—related or unrelated to string theory. Among the many advances we can expect, heavy ion collisions at the LHC will produce heavy quarks in much greater abundance than the ones at RHIC. Moreover, the detectors at the LHC are more advanced than the ones at RHIC. So it's reasonable to hope that greater clarity on the physical picture of energy loss from fast-moving quarks will emerge from the LHC.

It is fair to say, though, that the main suspense surrounding the LHC, is: What new particles will it discover? What new symmetries? Proton–proton collisions are far and away a better setting for such discoveries than heavy ion collisions, both because the energies per proton are higher and because the environment is less noisy. Prognosticating LHC discoveries is, naturally, more than a hobby among theorists. By the time you read this book, you probably will know more than I do now. But I'll hazard this guess: unless we're lucky, the discoveries won't flash like lightning bolts across the sky. The experiments are hard, the theories are abstract, and matching the two may well involve difficulties and controversies that are sharper than the ones I've described in this chapter. Even if a few discoveries are made right away, fitting everything into a coherent picture is probably going to be a long and confusing process. Because of its achievements to date, because of its rich mathematical structure, and because of the way it twines through such a broad swath of other theoretical ideas, from quantum mechanics to gauge theory to gravity, I expect that string theory will be a crucial part of the final answer.

CHAPTER EIGHT

EPILOGUE

THERE ARE MANY ASPECTS OF STRING THEORY WE COULD
ponder after the tour of the subject we have just completed.
We could ponder the peculiar demands it seems to make
on spacetime, like ten dimensions and supersymmetry. We
could ponder the peculiar objects whose existence it requires,
everything from D0-branes to end-of-the-world branes. We
could ponder its tenuous but improving connection to experimental physics. We could also weigh the controversy that
it generates: Is string theory worthwhile? Overhyped? Un-
duly maligned?

As fascinating as all these topics are, the topic that I think
is most worth ending with is the mathematics that forms
the core of string theory. Readers of my generation may
remember the Wendy's commercial where the short grey-
haired lady demands, "Where's the beef?" Well, in string
theory, the beef is in the equations. Almost all the equations
of string theory involve calculus, which puts them out of
reach of a popular account. So I've tried to take a handful
of important equations, roughly following the topics we've
covered in chapters 5 through 8, and put them into words.

The most basic formula in string theory is the equation
for how strings prefer to move. What this equation says is
that strings try to move through spacetime in such a way
that the area of the surface they sweep out is minimized.

160

This motion takes no account of quantum mechanics. There is another equation—really a group of equations— that explains how to include quantum mechanics in the motion of strings. These equations say that any motion of the string is possible, but that motions that are only slightly different from the area-minimizing motion reinforce one another. What I mean by "reinforcing" is illustrated by a Roman fasces: a bundle of sticks all lined up with one another. Such a bundle is very strong, much stronger than each stick. Each possible motion of a string is like a single stick. Most are scattered in a disorganized way. But motions of a string that are close to the area-minimizing motion are "aligned" in a way that makes them dominate the equations describing the quantum mechanics of strings.

The equations describing D-branes are variants of the ones that describe strings. Their most distinctive feature is that when many D-branes are grouped close together (again like a Roman fasces), they have more ways of moving than there are dimensions of spacetime. Whenever the D-branes separate significantly, ten-dimensional spacetime describes their relative positions. But when the D-branes are close enough, it takes a gauge theory to describe their motion. The equations of gauge theory say that strings stretching between pairs of branes, like I drew on p. 90, can't reliably be said to go from a "red" brane to a "blue" brane, or from "green" to "red." Instead, all such possibilities can be super-posed in a single colorful wave function, the way melody and harmony in the Fantasie-Impromptu combine without losing their separate identities.

The equations of string dualities have a remarkably frag-mentary quality. The ones that enter at the level of supergravity are surprisingly simple, usually expressing some symmetry relation. The ones describing strings and branes are quantum mechanical, but still pretty simple: the most common sort of equation in this context says that the electric charges (or analogs of electric charges) of branes must take on integer values in appropriate units. There are many, many further equations in string dualities, commonly arising from carefully tracing out how the intuitive relations we have discussed can be rendered quantitative. An example is the calculation of how the quantum fluctuations of a clump of D0-branes contribute to the mass of the clump. The answer—that they don't contribute at all—was anticipated based on the duality with M-theory long before it was demonstrated conclusively with equations.

The equations of supersymmetry start with relations like $a \times a = 0$. This equation has several meanings. It means that there are only two states of motion in a fermionic dimension: moving or stopped. It also means that two fermions can't occupy the same state (the exclusion principle), as we discussed for electrons in a helium atom. Supersymmetry goes on from simple relations like $a \times a = 0$ to truly profound equations that have helped shape modern mathematics.

The equations describing black holes and the gauge/string duality come mostly in two varieties. The first sort of equation is a differential equation. These equations describe the detailed moment-to-moment behavior of a string or a particle in spacetime, or of spacetime itself. The second sort of equation has a much more global flavor. You describe what is happening in a whole swath of spacetime, all in one go. The two types of equations are often closely related. For example, there's a differential equation that basically amounts to a particle saying, "I'm falling!" And there's a global equation describing a black hole horizon that basically says, "Fall across this line and you'll never make it back out."

EPILOGUE

As important as mathematics is to string theory, it would be a mistake to regard string theory as just a big collection of equations. Equations are like brushstrokes in a painting. Without brushstrokes there would be no painting, but a painting is more than a big collection of brushstrokes. Without a doubt, string theory is an unfinished canvas. The big question is, when the blanks get filled in, will the resulting picture reveal the world?

EPILOGUE

INDEX

INDEX

148; quantum chromodynam-
ics (QCD) and, 148–49, 151–57;
quark-gluon plasma (QGP) and,
5–6, 141, 145–55
fission, 16–17
FitzGerald, George, 6
FOIL expansion, 123–24
frequency: black holes and, 47; $E = h\nu$
and, 27–28; FM radio waves and,
29; hertz (Hz) measurement of,
25–26, 29; light and, 31; overtones
and, 26, 55–58, 94, 126; photoelectric effect and, 29–31; piano
strings and, 25–27, 46, 55–57, 60,
85; pitch and, 55; Planck's constant and, 27; quantum mechanics
and, 20–21, 31; vibrating string
and, 25 (*see also* vibrations); zero-
point energy and, 45–48, 57–61

galaxies, 36–37
gauge/string duality, 82–84, 95,
110–16, 126, 141, 148–58, 160–61
gauge symmetry, 31, 160; axis of
rotation and, 78–81, 110, 121;
branes and, 79–84, 95; charge
and, 80–84, 111; fifth dimension
and, 140, 148–58; gravity and,
110–16, 158, 160; photons and, 80;
QCD string and, 153; quantum
chromodynamics (QCD) and,
111–14, 126, 151–57; spin and,
79–80 (*see also* spin); supersymmetry and, 126; turntable analogy
and, 80–81
general relativity, 104; branes and, 71,
97; energy and, 2–8, 17; gravity
and, 2, 17, 38, 42, 54; maximal
supergravity and, 54; reconciling
with quantum mechanics, 51–55;
solitons and, 96–97, 97, 100–3,
132; spacetime and, 17; string theory and, 17–18, 51, 54–55, 67–68;

supersymmetry and, 126–27; time
dilation and, 12–15, 32, 42–44
Geneva, Switzerland, 5
Global Positioning System (GPS),
43–44, 63–65
gluinos, 135–36
gluons, 5–6, 98, 111, 133, 143, 146;
branes and, 84, 90; cascade inter-
ruption and, 143; color space and,
113; confinement and, 143; fifth
dimension and, 141–55; gauge/
string duality and, 111–15; quantum chromodynamics (QCD)
and, 111–15; quark-gluon plasma
(QGP) and, 5–6, 141, 145–55;
renormalization and, 143
gold nuclei, 17, 144–45, 155
gravitational collapse, 47
gravitational fields, 52
gravitinos, 135
gravitons: attempts to detect, 41–42,
89, 92, 98; as bosons, 122; daughter, 52; energy and, 41, 51–54,
60, 89, 98, 111–12, 120–22, 132;
fermionic dimensions and, 121–27;
mass and, 52; mother, 52; renormalization and, 54; response to
electric charge and, 52; spin and,
121–22; splitting of, 54–55; string
dualities and, 111–12; string theory
and, 51–54, 60–61; supersymmetry and, 89, 120–22, 132; tachyons
and, 60–61; uncertainty principle
and, 52
gravity: anomalies and, 2; black holes
and, 34–48; compactification and,
70; D-branes and, 86; fundamental physics and, 4; gauge symmetry and, 110–16, 158, 160; general
relativity and, 2, 17, 38, 42, 54;
low-energy limit and, 71, 73, 94,
141; mass and, 12; Newtonian
theory of, 42; quantum mechanics and, 51–55, 85; redshift and,

INDEX

INDEX

INDEX

INDEX

Gubser, Steven Scott,
1972-

The little book of
string theory.